Understanding GIS
An ArcGIS® Pro Project Workbook
THIRD EDITION

David Smith
Nathan Strout
Christian Harder
Steven Moore
Tim Ormsby
Thomas Balstrøm

Esri Press
REDLANDS | CALIFORNIA

Cover imagery credits: USGS, FAO, NPS, EPA, Esri, DeLorme, TomTom, and other suppliers; topographic basemap—Esri, HERE, DeLorme, Intermap, increment P Corp., GEBCO, USGS, FAO, NPS, NRCAN, GeoBase, IGN, Kadaster NL, Ordnance Survey, Esri Japan, METI, Esri China (Hong Kong), swisstopo, MapmyIndia, © OpenStreetMap contributors, and the GIS User Community.

Esri Press, 380 New York Street, Redlands, California 92373-8100
Copyright 2017 Esri
All rights reserved. First edition 2011; Second Edition 2013.
21 20 19 18 17   1 2 3 4 5 6 7 8 9 10
Printed in the United States of America

Library of Congress Cataloging-in-Publication Data

Names: Smith, David, 1959 July 5- author. | Strout, Nathan, author. | Harder,
  Christian, author. | Moore, Steven D., 1956- author. | Ormsby, Tim,
  author. | Balstrøm, Thomas, author.
Title: Understanding GIS : an ArcGIS Pro project workbook / David Smith,
  Nathan Strout, Christian Harder, Steven D. Moore, Tim Ormsby, Thomas
  Balstrøm.
Description: Third Edition. | Redlands, California : Esri Press, [2017] |
  "Second edition 2013"--T.p. verso. | Distributed to the trade in North
  America by Ingram Publisher Services. | Includes index.
Identifiers: LCCN 2017004606 | ISBN 9781589484832 (Paperback : alk. paper)
Subjects: LCSH: ArcGIS. | Geographic information systems.
Classification: LCC G70.212 .H358 2017 | DDC 910.285--dc23 LC record available at https://urldefense.proofpoint.com/v2/
url?u=https-3A__lccn.loc.gov_2017004606&d=DwIF-g&c=n6-cguzQvX_tUIrZOS_4Og&r=RhmcbAxStnbJpr06ef1onND
eVX-gjVopdqeQ8i7DbIY&m=Qbqt540GuwL_UIDhFFE2ccQBrwfuPjbA-6LT4UlndQI&s=OUapWxcN87cDNZdz30uM6nc
zGD_DW8acu4f_oUJ9Aa4&e=

The information contained in this document is the exclusive property of Esri unless otherwise noted. This work is protected under United States copyright law and the copyright laws of the given countries of origin and applicable international laws, treaties, and/or conventions. No part of this work may be reproduced or transmitted in any form or by any means, electronic or mechanical, including photocopying or recording, or by any information storage or retrieval system, except as expressly permitted in writing by Esri. All requests should be sent to Attention: Contracts and Legal Services Manager, Esri, 380 New York Street, Redlands, California 92373-8100, USA.

The information contained in this document is subject to change without notice.

US Government Restricted/Limited Rights: Any software, documentation, and/or data delivered hereunder is subject to the terms of the License Agreement. The commercial license rights in the License Agreement strictly govern Licensee's use, reproduction, or disclosure of the software, data, and documentation. In no event shall the US Government acquire greater than RESTRICTED/LIMITED RIGHTS. At a minimum, use, duplication, or disclosure by the US Government is subject to restrictions as set forth in FAR §52.227-14 Alternates I, II, and III (DEC 2007); FAR §52.227-19(b) (DEC 2007) and/or FAR §12.211/12.212 (Commercial Technical Data/Computer Software); and DFARS §252.227-7015 (DEC 2011) (Technical Data – Commercial Items) and/or DFARS §227.7202 (Commercial Computer Software and Commercial Computer Software Documentation), as applicable. Contractor/Manufacturer is Esri, 380 New York Street, Redlands, CA 92373-8100, USA.

@esri.com, 3D Analyst, ACORN, Address Coder, ADF, AML, ArcAtlas, ArcCAD, ArcCatalog, ArcCOGO, ArcData, ArcDoc, ArcEdit, ArcEditor, ArcEurope, ArcExplorer, ArcExpress, ArcGIS, arcgis.com, ArcGlobe, ArcGrid, ArcIMS, ARC/INFO, ArcInfo, ArcInfo Librarian, ArcLessons, ArcLocation, ArcLogistics, ArcMap, ArcNetwork, *ArcNews*, ArcObjects, ArcOpen, ArcPad, ArcPlot, ArcPress, ArcPy, ArcReader, ArcScan, ArcScene, ArcSchool, ArcScripts, ArcSDE, ArcSdl, ArcSketch, ArcStorm, ArcSurvey, ArcTIN, ArcToolbox, ArcTools, ArcUSA, *ArcUser*, ArcView, ArcVoyager, *ArcWatch*, ArcWeb, ArcWorld, ArcXML, Atlas GIS, AtlasWare, Avenue, BAO, Business Analyst, Business Analyst Online, BusinessMAP, CityEngine, CommunityInfo, Database Integrator, DBI Kit, EDN, Esri, esri.com, Esri—Team GIS, Esri—*The GIS Company*, Esri—The GIS People, Esri—The GIS Software Leader, FormEdit, GeoCollector, Geographic Design System, Geography Matters, Geography Network, geographynetwork.com, Geoloqi, Geotrigger, GIS by Esri, gis.com, GISData Server, GIS Day, gisday.com, GIS for Everyone, JTX, MapIt, Maplex, MapObjects, MapStudio, ModelBuilder, MOLE, MPS—Atlas, PLTS, Rent-a-Tech, SDE, SML, Sourcebook•America, SpatiaLABS, Spatial Database Engine, StreetMap, Tapestry, the ARC/INFO logo, the ArcGIS Explorer logo, the ArcGIS logo, the ArcPad logo, the Esri globe logo, the Esri Press logo, The Geographic Advantage, The Geographic Approach, the GIS Day logo, the MapIt logo, The World's Leading Desktop GIS, *Water Writes*, and Your Personal Geographic Information System are trademarks, service marks, or registered marks of Esri in the United States, the European Community, or certain other jurisdictions. CityEngine is a registered trademark of Procedural AG and is distributed under license by Esri. Other companies and products or services mentioned herein may be trademarks, service marks, or registered marks of their respective mark owners.

Ask for Esri Press titles at your local bookstore or order by calling 1-800-447-9778. You can also shop online at www.esri.com/esripress. Outside the United States, contact your local Esri distributor or shop online at eurospanbookstore.com/esri.

Esri Press titles are distributed to the trade by the following:

In North America: *In the United Kingdom, Europe, the Middle East and Africa, Asia, and Australia:*

Ingram Publisher Services   Eurospan Group   Telephone 44(0) 1767 604972
Toll-free telephone: 800-648-3104   3 Henrietta Street   Fax: 44(0) 1767 6016-40
Toll-free fax: 800-838-1149   London WC2E 8LU   E-mail:eurospan@turpin-distribution.com
E-mail: customerservice@ingrampublisherservices.com   United Kingdom

Contents

Foreword   vii

Preface   ix

Lesson 1   Frame the problem and explore the study area   1
   Exercise 1a: Explore the study area   7
   Exercise 1b: Do exploratory data analysis   36

Lesson 2   Preview the data   59
   Exercise 2a: List the data requirements   59
   Exercise 2b: Examine the data   64
   Exercise 2c: Reframe the problem statement   76

Lesson 3   Choose the data   87
   Exercise 3a: Choose the datasets   89
   Exercise 3b: Choose a coordinate system   106

Lesson 4   Build the database   127
   Exercise 4a: Project a shapefile   130
   Exercise 4b: Copy a feature class   138
   Exercise 4c: Prepare the city and county data   141
   Exercise 4d: Prepare the river data   151
   Exercise 4e: Prepare the park data   153
   Exercise 4f: Prepare the block group data   161
   Exercise 4g: Prepare the parcel data   172

Lesson 5   Edit the data   183
   Exercise 5a: Edit a feature   183
   Exercise 5b: Create a new park feature   193

Contents continued

Lesson 6 Conduct the analysis 203
Exercise 6a: Establish proximity zones 206
Exercise 6b: Apply demographic constraints 216
Exercise 6c: Select suitable parcels 226
Exercise 6d: Clean up the map and geodatabase 241
Exercise 6e: Evaluate your results 249

Lesson 7 Automate the analysis 257
Exercise 7a: Set up the model 259
Exercise 7b: Build the model (part 1) 263
Exercise 7c: Build the model (part 2) 275
Exercise 7d: Run the model as a tool 283

Lesson 8 Present your analysis results 295
Exercise 8a: Create the main map 298
Exercise 8b: Create a layout using the LA River and inset maps 308
Exercise 8c: Finish the map 322

Lesson 9 Share your results online 335

Appendix Data and image credits 339

Index 343

Foreword

More than 20 years ago, I was involved with a book project here at Esri®. We produced and published a software workbook called *Understanding GIS: The ARC/INFO Method*. We were just a group of people at this little GIS software company in Redlands, California, who saw that the users of our ARC/INFO software needed better guidance on how to use these tools for real analysis projects. Our original goal had been to write a book called *Getting Started with ARC/INFO*. More than 500 pages later, we realized that we had described a methodology for doing GIS projects in their entirety and needed to change the title.

The key idea was to set up a problem in the book, provide the data, and then let the students work through the whole process. Little did we know that our book would become a worldwide best seller and inspire a generation of technically savvy geographers.

Fast-forward to 2017. GIS has evolved to keep pace with a changing technology landscape. There have been huge innovations in geospatial data models and user interfaces; GIS now lives and breathes on the web. Much has changed. Yet the need for our users to understand how to organize and think about a GIS project persists. This book, *Understanding GIS: An ArcGIS® Pro Project Workbook*, now in its third edition since 2011, carries on the spirit and method of the original that inspired it. I'm especially happy that this new edition—featuring the latest version of our next-generation ArcGIS® Pro software—was voluntarily undertaken (and executed with precision) by our friends at the University of Redlands, the college geographically closest to Esri headquarters. I'm pleased to say that the new edition carries the torch while also integrating the new paradigm. It is my hope that it will inspire yet another generation of GIS professionals and practitioners.

Clint Brown
Director of Software Products
Esri
Redlands, California

Preface

What's new in the third edition?
This third edition of *Understanding GIS: An ArcGIS® Pro Project Workbook* has been completely revised and tested to be compatible with ArcGIS® Pro 1.4. New graphics have been created using the Windows 10 operating system. Some steps have changed to reflect changes in how the software works. Lesson 9 (available online, as was the case in the first and second editions) has been completely rewritten to reflect the latest advances in the rapidly evolving ArcGIS℠ Online environment. The demographic data has been updated with 2015 census data. Parcel data has been updated to 2016. The final results of the analysis are updated to an entirely new set of park sites. Numerous small mistakes that came to our attention have been corrected.

Background
In 1990, Esri published a software workbook called *Understanding GIS: The ARC/INFO Method*. This book was the first to offer a practical, project-oriented introduction to a commercial GIS software product, and it became popular as both a self-study tutorial for working professionals and a lab manual in college classrooms.

ArcGIS Pro is different from the ARC/INFO software for which the original book was written. Computers and operating systems are different, too. Ongoing development by Esri has extended the software's capabilities and reworked its architecture to keep pace with advances in technology—not the least of which is the sharing of geographic data and applications on the web. When the original book was published, the World Wide Web was still a couple of years in the future. In addition to new GIS functionality, significant changes have also been made to GIS data models, data storage, and user interfaces. All these changes notwithstanding, the underlying geographic approach (what was then called "the ARC/INFO method") remains basically the same. That approach can be summarized as follows: frame the problem, explore the study area, prepare the data, perform the analysis, and present your results. The same method is followed in this new book.

This book isn't a new edition of the original—the original is part of software history. This is a new book, from start to finish. It draws inspiration from the original but incorporates the latest developments to the ArcGIS platform and reflects the world in which GIS is practiced today.

Description
Understanding GIS: An ArcGIS® Pro Project Workbook is a tutorial designed around a multifaceted problem: finding a suitable location for a park next to the Los Angeles River in Los Angeles, California. Using real spatial data (from the City of Los Angeles and other providers) and realistic requirements, you'll complete all the essential phases of a GIS analysis project, from planning to execution to follow-up.

Goals

One goal of this book is to help you become a proficient user of ArcGIS Pro software. Another goal, ultimately more important, is to teach the geographic approach to problem solving. GIS students are often frustrated because software tools are presented in contexts that don't make clear how the tools relate to one another or how they serve larger purposes. This book incorporates software functionality in a meaningful process of analytic thinking.

Of course, GIS has many uses that are not analytical—data management and cartography, to name a couple. Even within the general category of analysis, different kinds of problems require different tools and strategies. This book is not all things GIS. It is not all things ArcGIS, either. Many aspects of the software are not relevant to our problem and are not explored in the book. Nevertheless, the book covers a lot of ground. When you're finished, you should have a good working knowledge of the software and a strong sense of how to use it productively on your own.

Audience

People come to GIS with varying background knowledge and experience. We loosely classify this book as intended for "ambitious beginners." If you have no prior experience with GIS or any of its wellsprings (geography, cartography, earth science, and computer database technology among them), you may find it more challenging than someone with exposure to these areas. On the other hand, thanks to GPS and household Internet mapping apps, location-based technology is becoming ever more familiar. Many things that needed explanation a few years ago have now merged into the background of common technical savvy.

This book is mainly written for the following:

- College-level or graduate students in GIS-related disciplines who want to learn ArcGIS Pro software and GIS best practices
- College and university instructors who want a classroom lab manual to supplement their GIS instruction
- New GIS professionals who want to strengthen or expand their knowledge of ArcGIS Pro software
- Professionals in other technical fields who want cross-training in GIS
- Current ArcGIS Desktop users transitioning from earlier software versions to ArcGIS Pro

There are no prerequisites for the book, and we've tried to write it so it will work for anyone with a serious interest in learning GIS. During the course of development and review, we found that many experienced users found things of value in it, too. If you're brand-new to GIS and are using the book outside a classroom or support group, be sure to take full advantage of the book's online resources, at esri.com/Understanding-GIS-3.

Structure

As mentioned, the book follows a single project, so it's best to work through it from start to finish. Lesson results are provided on the book resource web page in the event that you can't do every lesson or run into trouble.

The book has nine lessons, each divided into two or more exercises of varying length. Lesson 9 is only introduced in the book: the exercises for this lesson must be downloaded from the book resource web page.

Exercises are fully scripted so you won't encounter gaps in instruction, but frequently repeated operations aren't spelled out in detail every time. For example, after you've clicked a ribbon button, we won't keep showing pictures of it. Likewise, after you've opened a pane a few times, we won't keep telling you how to open it.

Conceptual information is supplied as needed, either directly in the text or in callout boxes. A few complex topics have a full sidebar to themselves. Figures, mostly screen captures, are common throughout the book. Their main purpose is to show you the correct state of your software at a particular point. Some figures are annotated to emphasize settings that need attention or help interpret a result.

The appendix is the book's data and image credits. The index will help you find your way back to important concepts, operations, and tools.

The exercises were created on computers running the Windows 10 operating system. They should be compatible on other Windows operating systems, although there may be small differences in operating system paths and the like. All images that are screen captures of software reflect a Windows 10 default theme.

Resources

Online resource page. The book's online resources are on the book's resource web page, at esri.com/Understanding-GIS-3.

Go here to access the data, get lesson results, download lesson 9, and obtain a 180-day trial of ArcGIS Desktop software.

Lesson results. The book's exercises are cumulative, with the results of one exercise defining the starting point of the next. For this reason, your results at the end of every lesson must be correct. The lessons include many screen captures as visual confirmations of progress, so by the end of an exercise, you should know whether you got the correct results. If you did, you can carry them forward. If you have problems, or if you skip an exercise, you can get the resulting project files (containing data, maps, and layouts) for any lesson. Results are available on the book's resource web page.

Additional exercises. The exercise instructions for lesson 9, "Share your results online," must be downloaded from the *Understanding GIS* online resource page.

Other resources. The ArcGIS Desktop Help system, installed with the software and accessible online, provides comprehensive descriptions of software concepts and tools. Additional online resources such as blogs, forums, map galleries, videos, user communities, and access to technical support and training can be found at this website: http://resources.arcgis.com.

Disclaimer

The data used in this book is real. So are the efforts of the City of Los Angeles and several interest groups to improve the environmental quality of the Los Angeles River and its surroundings. The GIS project in this book, however, was developed entirely at Esri. For the sake of a story, we pretend that the project was sponsored by the Los Angeles City Council. In fact, neither the book nor the project have any affiliation with the city beyond permission to use its GIS data. Likewise, there is no affiliation with the Los Angeles River Revitalization Corporation or with any Los Angeles River advocacy organization.

Acknowledgments

This book would not have been possible without the cooperation of the City of Los Angeles Department of Public Works and the Bureau of Engineering. Special thanks to Randy Price and Ann-Kristin Karling of the bureau's Mapping Division, and to City Engineer Gary Lee Moore, for giving us access to the city's data. We thank the City of Los Angeles Department of Recreation and Parks for providing its parks data. We note that land parcels and attributes are maintained by the Los Angeles County Assessor's Office.

The idea for the GIS project in this book was inspired by the Los Angeles River Revitalization Master Plan (available at http://www.lariver.org). Images from the master plan are used in lesson 1 by permission.

Many Esri employees reviewed the book in whole or in part, tested exercises, and gave advice or help. Thanks to all of them for their skill and dedication.

Thanks to the University of Redlands students who tested and provided valuable feedback on the exercises for this third edition. We want to acknowledge the support of the University of Redlands and the Center for Spatial Studies as well as Center for Spatial Studies GIS interns Anyssa Haberkorn, Adrian Laufer, and Jack Hewitt. Special thanks to the students of SPA 110—Introduction to Spatial Analysis and GIS, from the fall 2015 through spring 2017 semesters, who provided valuable constructive feedback on the workbook lessons.

And, most of all, a huge thanks to Clint Brown of Esri for supporting this project through thick and thin.

Technical requirements

For this book, your computer must have the following components:

- Windows 8.1 or 10 operating system (with the most recent service packs).
- 64-bit, hyperthreaded dual core (recommend quad core) with 4 GB of memory and 4 GB of disk space, or better.
- DirectX11 (OpenGL 3.3) compatible graphics card with 2 GB RAM, or better.
- Microsoft .NET Framework 4.6.1 or later must be installed prior to installing ArcGIS Pro.
- Microsoft Internet Explorer 10 or 11 must be installed prior to installing ArcGIS Pro.
- ArcGIS Pro software, Advanced license level (provided).
- Additional 800 MB free disk space for installing the exercise data (exercise data provided).
- An Internet connection.

See http://pro.arcgis.com/en/pro-app/get-started/arcgis-pro-system-requirements.htm for updated system requirements.

Installing the 180-day trial of ArcGIS Desktop software

Completion of the exercises in this book requires ArcGIS Pro 1.4 or later software at the Advanced license level. If you don't have the software, you can get a 180-day trial on the *Understanding GIS* online resource page. Follow the on-screen instructions to complete the software installation. You will be prompted to enter the EVA code printed inside the back cover of this book. The trial is a fully functional, but nonrenewable, 180-day license.

Required software version

This book requires ArcGIS Pro 1.4 or later software with an Advanced license. Earlier software versions may not be fully compatible with exercise data and do not operate as described in the exercises.

Installing the project data

To install the exercise data, go to the link on the book resource web page.

You will need to read and accept the license agreement terms.

Accept the default installation folder or navigate to the drive or folder location where you want to install the data. The exercise data will be installed on your computer in a folder called C:\UGIS (or \UGIS in the folder where you installed the exercise data).

Lesson

1 Frame the problem and explore the study area

THE VOLATILE LOS ANGELES RIVER

is the reason that America's second-largest city was founded in its present Southern California location by Spaniards in 1781. (The area was originally settled thousands of years earlier by the Gabrielino-Tongva Tribe, a California Indian tribe also known as the San Gabriel Band of Mission Indians.) Its water was tapped for drinking and irrigation, and a new city spread out from the river across the coastal plain. By the turn of the 20th century, the river was surrounded by a thriving urban center. Every few decades, raging floods would crest the banks at various points, submerging entire neighborhoods. After the historic floods of 1938 that claimed more than 100 lives and washed out bridges from Tujunga Wash to San Pedro (figure 1-1), city leaders had seen enough. By 1941, the US Army Corps of Engineers had begun to straighten, deepen, and reinforce the once wild waterway. Much of its length was eventually lined in concrete, and the river was more or less tamed.

Today, the City of Angels—home to nearly 4 million people—is a vibrant world center of business and culture. Running straight through the heart of the city, the Los Angeles River now serves as a flood control channel (figure 1-2). Sadly, this once bucolic waterway that was so instrumental to the formation of the city later became known as something ugly and marginal. Mile after mile of angled concrete appealed only to graffiti artists and filmmakers, and save for the occasional televised rescue of some hapless Angeleno swept away by a winter storm-fed torrent, the river remained a part of the city ignored by most. The negative perception has stuck with the neglected river for decades.

But in recent years, as the city has densified and much of Southern California's wild lands have been appropriated for development, new attention has focused on the river corridor and the scattered pockets of open space that line its length. Although it must always

Figure 1-1. The historic floods of 1938 washed out bridges up and down the Los Angeles River, including this one at Colfax Street and Vernon Boulevard.

Photo by Los Angeles Times staff photographer. Copyright © 1938 Los Angeles Times. Reprinted with permission.

Figure 1-2. The Los Angeles River now serves as a flood control channel though the river corridor is the focus of city regreening efforts.

serve its important flood control function, the river and adjacent lands are increasingly recognized as underutilized, providing opportunities for regreening and psychic restoration for people living in an overbuilt city. Adventuresome and resourceful citizens have discovered peaceful pockets of sanctuary along the river and made these places their own. A vital and concerned activist community has raised awareness of the river and pushed for its beautification and redevelopment.

In 2005, the city launched a major public works project focused on the human dimensions of the river. A landmark study, the Los Angeles River Revitalization Master Plan, demonstrated the significant potential of redevelopment to improve the quality of life for citizens living near the river corridor. Then mayor Antonio Villaraigosa said at the time, "We have an opportunity to create pocket parks and landscaped walkways ... to create places where children can play and adults can stroll."

According to Villaraigosa, "The plan provides a 25- to 50-year blueprint for transforming the city's 32-mile stretch of the river into an 'emerald necklace' of parks, walkways, and bike paths, as well as providing better connections to the neighboring communities, protecting wildlife, promoting the health of the river, and leveraging economic reinvestment."

Although the 2005 master plan identified some of the most obvious areas for large-scale regional redevelopment along the river, it stopped short of identifying smaller (and more affordable) neighborhood projects; that work would require a more involved study. With thousands of land parcels strung out along the river, identifying the best places for park development is like looking for a needle in a haystack. Many factors come into play, among them current land use, demographics, and accessibility.

In the years since the plan's completion, the city has created a website (figure 1-3) that encourages people to learn about (and participate in) the latest developments related to its landmark resource. The website, at www.lariver.org, contains links to many resources about the river and its watershed, including scientific studies and recreational opportunities. If city leaders can find the resources and a motivated citizenry keeps up the pressure, a renaissance will transform growing stretches along the river into real versions of the revitalization effort's artists' renderings (figure 1-4).

Figure 1.3. The Los Angeles River Revitalization website contains links to information about the river and its watershed.

Courtesy of the City of Los Angeles, Los Angeles River Revitalization Master Plan

Here's where you pick up the thread in this book. You'll use the city's real need for river redevelopment as a launching point for a park siting analysis using a geographic information system (GIS). A GIS is ideal for this type of decision-making because it allows you to analyze large amounts of data in a spatial context. In this book, you'll spend a lot of time with ArcGIS Pro software, and by the end you'll have completed a project from start to finish. Along the way, you'll gain an excellent grasp of what a GIS can do.

You'll be assuming the role of a GIS analyst for the City of Los Angeles. So what exactly does a GIS analyst do, and how is that job different from other jobs that also use GIS software? Table 1-1 defines some of the various roles that a typical GIS operation might establish to accomplish its work.

Figure 1.4. The Los Angeles River Revitalization website contains artists' renderings (A to C) of a rejuvenated river corridor.

\	Table 1-1. GIS roles
Editors	People who create, update, and correct spatial data and its attributes (statistical and descriptive information).
Cartographers	People who make and publish maps and solve information design problems.
Analysts	People who query and process geographic data to solve analytical problems.
Programmers	People who implement custom GIS functionality by developing scripts and applications for specific procedures.
Managers	People who oversee staffing and equipment, database design, workflow, new technology, and data acquisition.

The central work in this book is analytical. Your main focus will be on using ArcGIS tools and methods to find the most park-suitable land within a study area, but there is preparatory work to do before the analysis proper, and there are results to interpret and present afterward. This book has two goals. One is to present a comprehensive approach to geographic problem solving. We want to help you develop skills, habits, and ways of thinking that will be useful in projects other than this one. The second goal is to teach you how to use ArcGIS Pro software. These goals are mostly complementary. ArcGIS is a big system, however, and it wouldn't be realistic to try to cover all that it can do in a single book. Our principle has been to teach the software in the service of the project and not otherwise. You'll delve into many aspects of ArcGIS Pro—editing, modeling, and cartography, among them—but there are other aspects that we won't use, or will only touch upon lightly, because they aren't strictly relevant to our needs. We might say (with apologies to Waldo Tobler[1]) that everything in a GIS is related to everything else, but some things are more closely related to analysis than others.

Frame the problem

The first step in the geographic approach to problem solving is to frame the problem. What that means, first of all, is coming up with a short statement of what it is you want to accomplish. For this project, you want to find a suitable site for a park near the Los Angeles River.

Once you have the statement, you can begin to tease out its ambiguities. What factors make a site "suitable"? In this case, the city council has already established a concise and fairly specific set of guidelines.

---

1. Tobler, W. 1970. "A Computer Movie Simulating Urban Growth in the Detroit Region." *Economic Geography* 46 (2): 234–40.

Lesson One road map

You are here:

What you'll do in this lesson:

1. Frame the problem and explore the study area
   - 1a Explore the study area
   - 1b Do exploratory data analysis
2. Preview the data
3. Choose the data
4. Build the database
5. Edit the data
6. Conduct the analysis
7. Automate the analysis
8. Present your analysis results
9. Share your results online

Park guidelines
1. On a vacant parcel of land at least one-quarter acre in size
2. Within the Los Angeles city limits
3. As close as possible to the Los Angeles River
4. Not in the vicinity of an existing park
5. In a densely populated neighborhood with lots of children
6. In a lower-income neighborhood
7. Where as many people as possible can be served

This list limits the scope of your inquiry, but it's far from a complete breakdown of the problem. Some of the guidelines are specific, but others are vague. Familiar concepts are sometimes the hardest to pin down. For example, what income level should count as "lower income"? How are the boundaries of a "neighborhood" established? You can't solve the big problem until you've solved the little problems buried inside. Usually, however, it's not possible to address (or even foresee) all the little problems ahead of time.

Data exploration influences the framing of the problem. Do you have income data on hand? If so, is it for individuals or households? Is it average, median, or total? To the extent that the questions themselves are indefinite (What is lower income? How should it be measured?), the data you have available will help shape the answer.

Analysis also influences the framing of the problem. Given that you want a quarter-acre tract of vacant land, what do you do about adjacent lots? Is there a tool to combine them? If so, does using it have unanticipated side effects, such as a loss of information? Your data processing capabilities (and your knowledge of them) may determine how you define a "one-quarter acre parcel."

Even the results of an analysis influence the framing of the problem. Suppose, after having carefully defined the guidelines, you run the analysis and don't find any suitable sites. Do you report to the city council that there's just no room for a park? More likely, you'll change some of your definitions and run the analysis again.

Framing the problem, therefore, is an ongoing process, one that will occupy you throughout much of the book.

Exercise 1a: Explore the study area

In this exercise, you'll get to know the Los Angeles River and its surrounding area with maps and data. At the same time, you'll learn the basics of working with ArcGIS Pro: how to navigate a map, add and symbolize data, and get information about map features.

Start ArcGIS Pro

Now you'll open the ArcGIS Pro application.

▶ See the preface for how to install ArcGIS Pro.

1) **Start ArcGIS Pro by clicking the Start button on the taskbar, and then, on the Start menu, click All Programs > ArcGIS > ArcGIS Pro > ArcGIS Pro.**

To open ArcGIS Pro, you must sign in to ArcGIS Online using an organizational account. Signing in to ArcGIS Online allows you to access and share GIS content with users in your organization as well as publicly with users around the world.

2) **Click Sign In.**

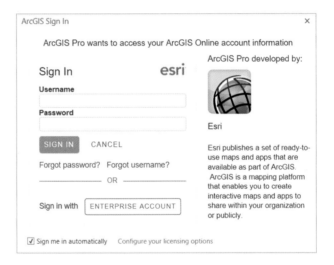

The application opens with the ArcGIS Pro dialog box.

Create a new project

1) **To create a new project, click Blank.**

2) Provide a New Project name: LARiver_ParkSite.

3) Browse to and select the UGIS (Understanding GIS) folder as the location to save the new project folder.

4) Click OK.

Creating the project creates a new folder in the UGIS folder with the name you gave it, along with some default items, including a new geodatabase and a toolbox.

Insert a new map

1) On the Insert tab, click New Map.

A new map is added to the project using the Topographic map by default.

On the right side of the application is the Project pane. This pane contains all the maps, data, tools, and other resources associated with the project.

2) Open the Project pane (if necessary) and expand the Maps item.

Note that a new map simply named Map has been added to your project.

▸ If the Project pane is not displayed, go to the View tab and click Project.

3) Rename the new map by right-clicking Map in the Project pane and clicking Rename. Type Lesson1a and press Enter.

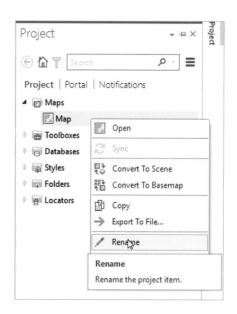

Change the basemap to Streets

1) On the Map tab, change the basemap by clicking the Basemap button ⊞ in the Layer group. The basemap gallery is displayed with several different basemap options.

2) Click the Streets basemap.

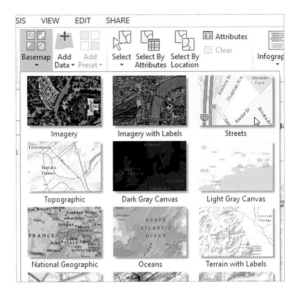

The World Street Map basemap layer is added to your Lesson1a map. It has an entry in the Contents pane on the left. Again, click the View tab, if necessary, to open the Contents pane.

The Map tab at the top of the application includes tools for navigation, layers, selections, inquiries, and labeling. As different tabs are selected—Insert, Analysis, View, Edit, and Share—different tools will appear on the ribbon. Additional contextual tabs, including Appearance, Labeling, and Data, will appear depending on the tasks that you are performing.

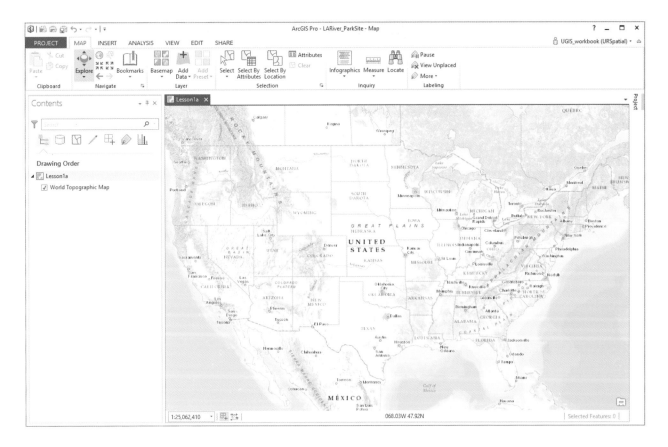

Zoom in to Southern California

Now you'll zoom in to your area of interest.

1) **Experiment with navigating the map with the mouse.**

   A. Push the scroll wheel forward to zoom in.

   B. Pull the scroll wheel back to zoom out.

   C. Drag in any direction with the left mouse button to pan.

   D. Drag up with the right mouse button to zoom out.

   E. Drag down with the right mouse button to zoom in.

   F. Press and hold the Shift key and drag a box with the left mouse button to zoom in.

   G. In the map window, use the Shift key to drag to draw a box around Southern California, as shown in the figure.

   Your box doesn't have to match exactly.

2) Zoom in again to get closer. When you see city names and major roads, pan (left mouse button) to center the view on the Los Angeles area.

3) Keep zooming in (try the Fixed Zoom In button, too) until you can easily distinguish cities, freeways, and landmarks such as airports.

4) If you zoom in further than you want, zoom out (pull the scroll wheel back), or click Fixed Zoom Out in the Navigate group to go back.

You probably noticed that no streets were visible at the global scale, and that as you kept zooming in, more and more detail appeared. This is because the basemap is a multiscale map—really, a set of maps that turn on and off to display features and symbology that are appropriate to your map scale.

The figure shows Greater Los Angeles, an area that includes scores of incorporated communities and nearby cities.

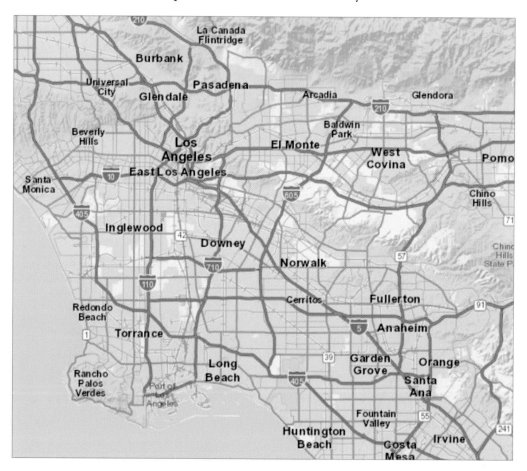

Add a layer of project data

On top of the basemap, you'll add a layer from the data that has been put together for this project and is stored on your computer. In lesson 2, we'll talk more about where this data comes from and how to acquire data of your own.

1) In the Project pane, right-click Folders and click Add Folder Connection.

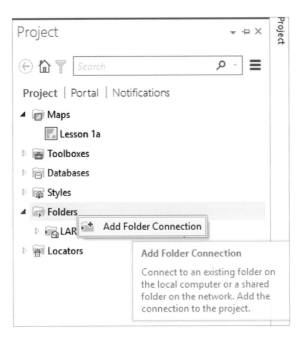

Basemap layers

Basemap layers show reference geography such as street maps, imagery, topography, and physical relief. The ones available in ArcGIS Pro are remotely hosted map services that you can navigate, view, and use as backdrops to other data. Basemaps are stored at multiple scales, so that as you zoom in or out, you see different amounts of detail. As you navigate a basemap, the various pieces of it (called *tiles*) that compose your current view are stored locally on your computer in a so-called "display cache." When you zoom or pan to a new area, the map may be a little slow to draw, but anyplace you return to will redraw quickly because the data comes from your cache, not from the remote server.

2) Browse to your UGIS folder and select the ParkSite folder. Then click OK.

Selecting a folder adds a new folder connection to ParkSite.

3) After making the folder connection, expand ParkSite\SourceData and then ESRI.gdb, and finally the Boundary item.

4) Drag and drop the City_ply layer on your map. Default layer colors vary so don't worry if the color of your layer looks different from the figure.

We'll discuss GIS data formats in the sidebar "Representing the real world as data" in lesson 2. For now, you just want to dig down to your data.

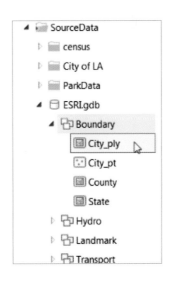

When the layer of city boundaries is added to the map, it may zoom to the full extent of the dataset. If so, click the Previous Extent button ← in the Navigate group on the Map tab.

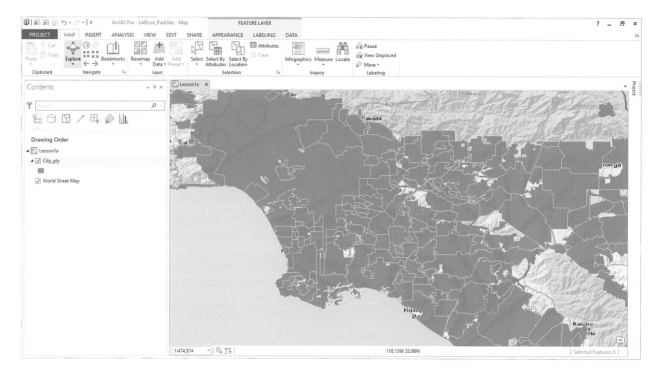

Each city in the layer is called a *feature*. These features are polygons, which are one of the three basic shapes used to represent geographic objects in a GIS. (The others are lines and points.)

In the Contents pane, the order of entries matches the drawing order of layers in the map. City_ply is listed above World Street Map, and on the map, the cities cover the basemap. You can control a layer's visibility with its check box in the Contents pane.

5) **In the Contents pane, click the check box next to City_ply.**

The layer is turned off.

6) **Click its check box again to turn the layer back on.**

Set layer properties

Every layer has properties you can set and change. For example, you just changed the visibility property of the City_ply layer.

1) **In the Contents pane, right-click the color patch underneath the City_ply layer name. A color palette opens. Moving the pointer over any color square shows its name as a ToolTip.**

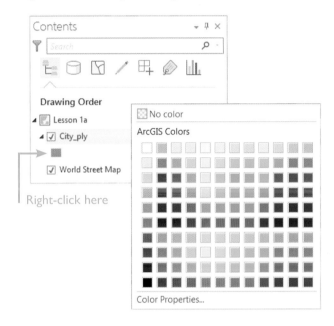

2) **On the color palette, click any color you like to change the layer color.**

On the map, the cities redraw in the color you chose.

3) **In the Contents pane, right-click the City_ply layer name to open its context menu. At the bottom of the menu, click Properties.**

The Layer Properties dialog box opens. Here is where you access the full set of properties for a layer.

4) **If necessary, click the General tab at the top of the left column on the dialog box.**

The layer's name, City_ply, is one of its properties. This name is cryptic (it stands for "city polygons") and unattractive, so now you can change it.

5) **In the Name box, delete the text and type Cities.**

6) **In the Layer Properties dialog box, click the Source tab.**

This tab shows technical information about the layer, including the path to the data on your computer.

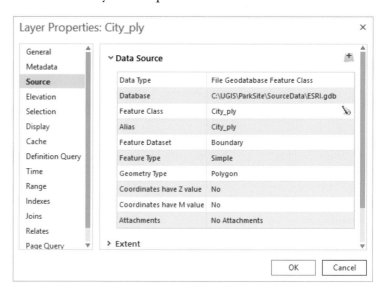

7) **Click OK on the Layer Properties dialog box to close it.**

Notice that the name is updated in the Contents pane.

Renaming the layer in the map simply makes it easier to identify in the map, it doesn't change the source data file's name (which is still City_ ply). A layer is a representation or rendering of the data, not the data itself. You can make any changes you want to a layer's properties without affecting the data on which the layer is based.

Get information about cities

Now see what you can find out about the cities on the map.

1) Click any city polygon on the map.

The city flashes blue, and a pop-up window opens. In the title bar of the pop-up window, you see the name of the city you identified. In the pop-up window, you see its attributes, or the information that this layer stores about cities. Some of the attributes aren't obviously meaningful, but others are. If POP2010 is population for the year 2010, and POP10_SQMI is population per square mile for the same year, you know that Los Angeles (if that's the city you identified) had 3,792,621 inhabitants at the time of the 2010 census, and its population density was 8,018 people per square mile.

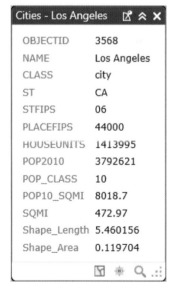

2) If necessary, move the pop-up window away from the map. Identify a few more cities.

The pop-up window updates with information about the new city. All the cities have the same set of attributes; it's the values of the attributes that change.

3) Close the pop-up window.

Exercise 1a: Explore the study area   17

4) In the Contents pane, right-click Cities and click Attribute Table.

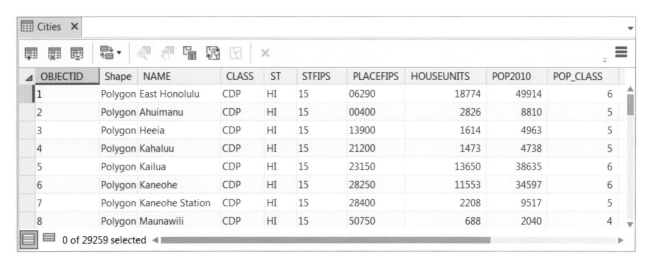

5) Scroll across the table, if necessary, to look at all the field names (the gray column headings).

This is a different presentation of the same information you saw when you identified cities on the map.

6) Scroll back to the beginning of the table, and then scroll a little way down through the records (the table rows).

This table has a lot of records: in fact, 29,259 of them, as you can see at the bottom. Each record corresponds to a unique feature—that is, a unique city—on the map. So there must be a lot of cities you're not seeing in the current view.

7) Leave the table open. It will stay docked under the Lesson1a map.

▶ You can move the panel by dragging it from the tab and redock it by dropping it on any blue arrow.

8) In the Contents pane, right-click the Cities layer. On the context menu, click Zoom To Layer.

The map zooms out to the geographic extent of the layer: the entire United States. You can't distinguish individual city polygons at this scale.

9) On the Map tab, in the Navigate group, click the Previous Extent button.

Select the record for Los Angeles

When you select a record in an attribute table, the corresponding feature is selected on the map. (Likewise, when you select a feature on the map, its record is selected in the table.) Selections are marked with a blue highlight by default.

1) Scroll up to the top of the table. Make sure the table is wide enough that you can see the POP2010 field.

▶ If necessary, widen the table by dragging its edge.

2) Right-click the POP2010 field name. On the context menu, click Sort Descending. Alternatively, double-click the column header to switch between the ascending/descending sort order.

The records are sorted in the order of their populations, from largest to smallest. Los Angeles is now the second record in the table, after New York.

3) At the left edge of the table, click the small gray box next to the Los Angeles record to select it.

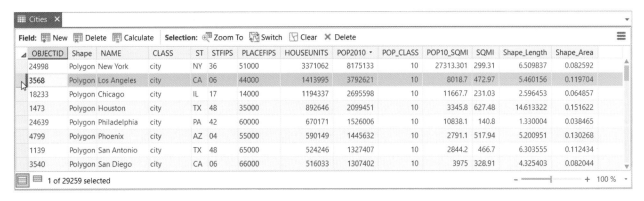

On the map, the city of Los Angeles is outlined in blue. You may be able to see the whole city already, but now you can make sure.

4) Close the attribute table by clicking the X on the right of Cities.

5) In the Contents pane, right-click Cities and click Selection > Zoom To Selection.

Exercise 1a: Explore the study area   19

The map zooms in close on the selected feature and centers it in the view. The city's odd shape is attributable to years of piecemeal expansion and incorporation. The city has internal "holes" where it surrounds other cities, such as Beverly Hills, or unincorporated areas. It also has a long, narrow southern corridor that connects it to its harbor at the Port of Los Angeles.

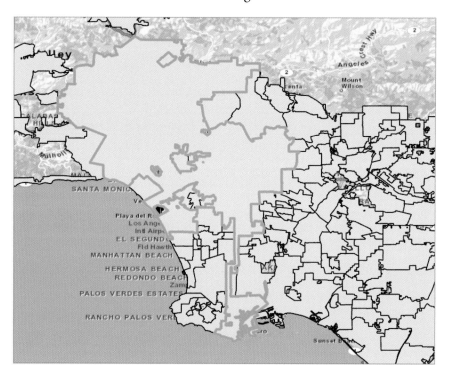

6) On the Map tab, in the Selection group, click the Clear button to unselect the feature.

Filter the display of cities using a definition query

One of your project requirements is that the new park be inside the Los Angeles city limits. Therefore, you're more interested in Los Angeles than in other cities. A layer property called a *definition query* allows you to show only those features in a layer that interest you.

1) In the Contents pane, right-click the Cities layer and click Properties.

2) In the Layer Properties dialog box, click Definition Query.

3) Click Add Clause to begin building your query.

You build a query on an attribute table by specifying a field and setting a logical or arithmetic condition that values in that field must satisfy. In this case, you want to find records with the value of Los Angeles in the NAME field.

4) In the list of field names, use the drop-down arrow to click NAME.

5) Click is Equal to in the second drop-down menu.

6) Type Los Angeles in the third box. You can also use the menu to select the city name.

7) Click Add. Do not click OK yet.

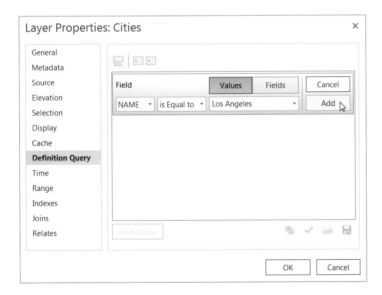

8) Click the General tab. Delete the layer name Cities and type Los Angeles.

9) Click OK.

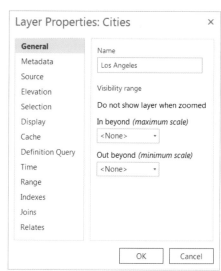

On the map, only the city of Los Angeles is shown. The other cities are hidden by the definition query.

22   Lesson 1: Frame the problem and explore the study area

10) **In the Contents pane, right-click the Los Angeles layer and click Attribute Table.**

The table shows only the record for Los Angeles.

11) **Close the attribute table.**

The other city features haven't been deleted. Clearing the definition query would display them again. Layer properties affect the display of data in a map, not the essential properties of the data itself: the number of features, their shapes, locations, and attributes.

Add a layer of rivers

Now you can add a layer of local rivers to the map and see where the Los Angeles River fits into the picture.

1) From the Project pane, browse to the Hydro group in the ESRI.gdb and right-click the River layer. Then click Add To Current Map from the menu.

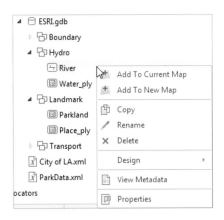

> **Layers and datasets**
>
> A layer points or refers to a dataset stored somewhere on disk (as specified on the Source tab of the Layer Properties dialog box). A layer is not a physical copy of the data. A layer is a representation or rendering of the data. The dataset stores the shapes, locations, and attributes of features; the layer stores display properties, including what the layer is named, which of its features are shown or selected, how those features are symbolized, and whether they are labeled. Changes to layer properties do not affect the dataset that the layer refers to. You can make as many layers as you want from the same dataset and give them different properties. These layers can co-exist in the same map document or in many different map documents.

A layer of rivers is added to the map and placed at the top of the Contents pane. The layer is symbolized with a random color; if the color is difficult to identify, it may be beneficial to change its color using the color palette (if you are having trouble, refer to the task "Set layer properties," steps 1 and 2, earlier in this section, for instructions).

2) **Click on any river to identify it.**

You see the name of the river (many of them don't have names) and its other attributes.

By default, the Explore tool identifies features from the topmost layer in the Contents pane. If you miss a river, you'll identify either the city of Los Angeles or nothing. That's fine—leave the Explore window open and click again on a river.

Exercise 1a: Explore the study area 23

3) Click on a few more rivers to identify them. Try to identify a segment of the Los Angeles River.

The river runs west to east across the city, turns south near the city's eastern edge, and follows a freeway to San Pedro Bay.

4) Close the pop-up window.

Make a definition query on the LA River

Just as you're mainly interested in one city, you're also mainly interested in one river. You'll make another definition query to show just the Los Angeles River.

1) In the Contents pane, right-click the River layer and click Properties.

   ▶ A shortcut is to double-click the layer name in the Contents pane.

2) In the Layer Properties: River dialog box, click Definition Query.

3) Build your query:

   Name is equal to Los Angeles River.

4) Click the General tab. Change the layer name from River to Los Angeles River.

5) Click OK.

The Contents pane reflects the new layer name.

   ▶ You can also rename a layer directly in the Contents pane by clicking the name once to highlight it and then clicking it again to rename it. (Be careful not to double-click, or you'll open the layer properties.)

Change the symbology

1) Click the color symbol below the Los Angeles River name to modify the symbol. Clicking the symbol opens the Symbology pane on the right side of the map.

2) Click the Properties tab at the top. Under Appearance, you can change the color and width of the symbol.

3) Change the color to blue.

4) Change the line width to 3 pt.

5) Click the Apply button and close the Symbology pane.

On the map, the river is displayed with its new symbology.

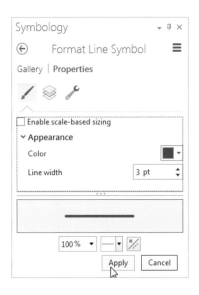

Select Los Angeles River features using a query

On the map, the river looks like a single feature (just like the city), but it's not.

1) In the Contents pane, right-click the Los Angeles River layer and click Attribute Table.

At the bottom of the table, you see that 0 of 17 records are selected. That means that, in this particular layer, the Los Angeles River is composed of 17 features. Why is that?

2) Scroll down through the table.

All the records have the same name value. Most have the same type, but one is an artificial path. There are a few description values. The need to maintain different attribute values for different parts of a geographic object is a common reason that data—especially data representing linear features such as streets and rivers—is constructed this way, consisting of multiple features. We'll come back to this point in lesson 2.

3) **In the ribbon of tools on the Map tab, in the Selection group, click the Select By Attributes button** .

Having noticed "perennial" (year-round water flow) and "intermittent" (not year-round water flow) values in the description field, you might want to know which parts of the river are which. You can find out using an attribute query. An attribute query is like a definition query in that both types of query single out features in a layer on the basis of attribute values. The difference is that an attribute query highlights (selects) features that satisfy the expression rather than hiding features that don't.

4) **In the Layer Name or Table View box, confirm that Los Angeles River is selected.**

5) **Make sure the Selection type drop-down list is set to New selection.**

6) **Click the Add Clause button.**

7) **Create a clause for**
    Description is equal to perennial.

8) **Click Add.**

9) **Click Run, in the lower-right corner of the pane.**

Twelve records are selected in the table. The corresponding features are selected in the map, showing that the river is perennial for most of its length.

It's not among your guidelines to locate the new park along the river's perennial stretch, but it's interesting that fairly simple data exploration may introduce new ways of thinking about your problem.

10) **Close the Select Layer By Attribute Geoprocessing pane.**

Find the length of the perennial portion of the LA River

As you learned by looking at the attribute table, the perennial portion of the LA River is separated into 12 distinct records. To find the total length of this section, you must perform a summation of the FEET fields, using the Summary Statistics tool. When features are selected in the layer, the Summary Statistics tool processes only the selected records as a subset. Because you have perennial streams selected, the tool will sum up only the SUM field of these selected features.

1) In the attribute table of the Los Angeles River layer, right-click the DESCRIPTION field and click Summarize.

2) Rename the output table LARiverPerennial by clicking the Browse button and entering the name.

3) In the Field drop-down menu, click FEET, and in the Statistic Type menu, click SUM.

So you're using the Summary Statistics tool to find the total length (sum of each FEET record) of the perennial portion of the LA River.

4) In the Case field drop-down menu, click DESCRIPTION.

5) Compare the tool to the figure and click Run in the lower-right corner of the Geoprocessing window.

6) When the tool finishes processing, the resulting table is added to the bottom of the Contents pane. Right-click the table and click Open to see the results. Confirm that the total feet of perennial streams is 229,947.

7) At the top of the Los Angeles River attribute table panel, click the Clear Selection button.

Clearing the record selection clears the feature selection on the map and in the table.

8) Close the attribute tables of Los Angeles River and LARiverPerennial so that the map fills the middle of the application. You can also close the Summary Statistics pane now.

9) Save the project by clicking the Save button at the top of the application.

Save your project often as you work through the workbook.

Selecting features

You make selections to work with a subset of features in a layer. You can use a selection to zoom the map to a specific area, make a new layer from the selected subset, get statistical information about the subset, or for many other things.

Selections are also used in queries. Whether you do an attribute query (to find records with a certain attribute value) or a spatial query (to find features with a certain spatial relationship to other features), the records and features that satisfy the query are returned as a selection on the layer.

Change the basemap to Imagery with Labels

Imagery provides a detailed, photorealistic view of the ground, and you'll rely on it to explore the LA River in more detail. Imagery also has other important uses, such as providing a background against which to edit features (lesson 5) and "ground truthing" analysis results (lesson 6).

1) On the Map tab, click the Basemap drop-down arrow and click Imagery with Labels.

World Imagery is added to the bottom of the Contents pane and replaces streets. A layer named World Boundaries and Places has also simultaneously been added to the top of the Contents pane. This layer consists of locations, boundaries, and labels and should be visible on the map. (You should see yellow place-names covering the city of Los Angeles feature, for instance). Your Contents pane and map should look like the figure.

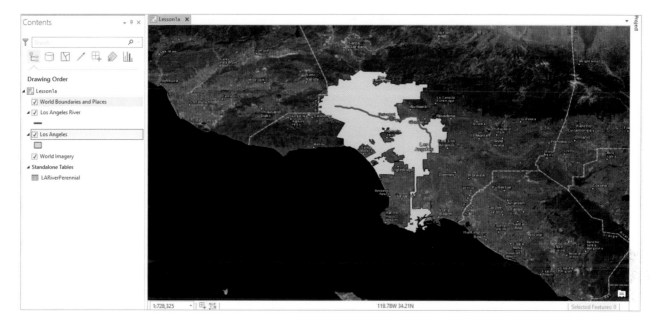

Create a bookmark

Certain views in a map are useful for orientation or reference. You can bookmark a view to make it easy to return to.

1) First, zoom to the Los Angeles layer by right-clicking the Los Angeles layer and clicking Zoom to Layer.

2) On the Map tab, click Bookmarks > New Bookmark.

3) In the Create Bookmark dialog box, replace the default name with City of Los Angeles. Click OK.

4) Zoom in on the river's mouth at the Port of Long Beach by pressing and holding the Shift key.

5) In the lower-left corner of the map, click the Map Scale drop-down arrow and click 1:10,000.

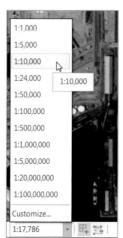

At this scale, one unit of measure on the map is equivalent to 10,000 of the same units on the ground. Loosely, a thing on the map is 10,000 times smaller than its actual size.

Exercise 1a: Explore the study area 29

6) Use the Zoom and Pan tools to explore the harbor.

The imagery is high resolution, and you'll see more detail as you keep zooming in. Eventually, you'll reach a limit in the level of detail, and the image will become pixelated.

7) When you're ready, on the Map tab, click Bookmarks > City of Los Angeles.

The map view returns to the bookmarked extent.

Change the symbology for Los Angeles

The boundary of Los Angeles is filled in with a solid color so the imagery underneath is still covered up. Soon you're going to zoom in and follow the river's course through the city. It will serve that purpose to resymbolize the city so that you see only its outline.

1) In the Contents pane, click on the color symbol below the name Los Angeles to modify the symbol.

The Symbology pane opens on the right side of the map.

2) Click Properties. If necessary, expand Appearance.

3) Change the color to No Color.

4) Change the outline color to Autunite Yellow.
5) Increase the outline width to 2 pt.
6) Click Apply.

The new symbology is displayed on the map.

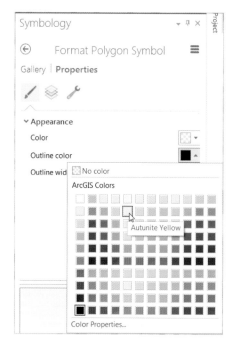

7) Close the Symbology pane.

Save the layer as a layer file

It wasn't hard to make this particular symbol, but it often takes time and effort to create good symbology. Having done so, you may want to reuse that symbology. In this book, for instance, you'll draw the outline of Los Angeles in other map documents in coming lessons.

You can save layer properties to a file called a *layer file*, which has the file extension .lyrx. A layer file is not a copy of the data, but you add it to a map document in the same way that you add data. The layer file stores all the properties of a layer—its name, symbology, definition query, and so on—including the path to the layer's source dataset. When you add a layer file to ArcGIS Pro, the layer draws with its properties already set.

Exercise 1a: Explore the study area   31

1) **In the Contents pane, right-click Los Angeles and click Save As Layer File (near the bottom of the context menu).**

The Save Layer(s) As LYRX File dialog box opens.

2) **Click Folders at the top of the left column and open the ParkSite\MapsAndMore folder.**

3) **Name the new file LosAngeles (no spaces) and click Save.**

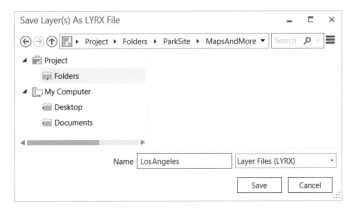

Now you'll remove the Los Angeles layer that's currently in the map, and then add the layer file to see how it works.

4) **In the Contents pane, right-click Los Angeles and click Remove.**

The layer disappears from the map and the Contents pane.

5) **In the Project pane, browse to the MapsAndMore folder.**

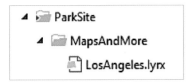

6) **Drag the LosAngeles.lyrx file to your map.**

The layer is added to the map with all its properties set (layer name, symbology, definition expression, and so on).

7) **Open the layer properties of Los Angeles. In the Layer Properties dialog box, click Source.**

Note that as before, the layer points to the City_ply feature class. As stated previously, layer files do not store raw data; they are a pointer to the GIS data along with the properties about how to display the data on a map.

8) Close the Layer Properties dialog box.

Anytime you add the LosAngeles.lyrx file to a map document, the layer will draw as a yellow outline with a definition query on the city of Los Angeles. Once the layer is added to the map, however, it's just the same as any other layer, and you can change its properties however you want. (Not that you want to change them.)

Follow the river

Now you can start developing a sense of the study area by following the river's course through the city.

1) Maximize the ArcGIS Pro window if you haven't done so already.
2) Open the City of Los Angeles bookmark.
3) Zoom in on the river's source in the community of Canoga Park in northwest Los Angeles.

The river officially starts where Bell Creek and the Arroyo Calabasas converge at Canoga Park High School.

4) Click the Map Scale drop-down arrow below the map and click 1:24,000.
5) Press and hold the left mouse button to pan slowly eastward along the river.

Densely populated residential neighborhoods line both sides of the river until you get to the Sepulveda Dam Recreation Area, a large recreational area with golf courses and a lake. The river bottom is natural here, becoming concrete again at the Sepulveda Dam in the southeastern corner of the basin.

> **Setting map scale**
>
> By default, map scale is displayed in the form of a representative fraction, such as 1:24,000. At this scale, for any unit of measure, one unit of distance on the map is equivalent to 24,000 units in the real world. You can set the map to any scale you want by typing a number in the scale box and pressing Enter. You can also enter a verbal expression such as "1 inch = 1 mile," and it will be converted to a representative fraction. To change the way that scale is displayed, or to change the list of predefined scales, click Customize This List at the bottom of the Map Scale drop-down list.

East of the Sepulveda Basin, the river follows a freeway for a while and is again surrounded by fairly dense residential and commercial areas.

6) **On the keyboard, press and hold the Q key.**

Now you roam continuously across the display in whichever direction you point the mouse. To control your speed, make small brushing movements with the mouse either with or against the direction of movement. As you roam, the imagery should draw smoothly and continuously, although your experience may vary. Other layers, such as the Los Angeles River layer, suspend drawing and catch up when you stop.

7) **Release the Q key to stop roaming.**

▶ You can also use the four arrow keys on the keyboard to roam.

8) **Continue to pan (or roam) along the river.**

The river flows generally southeast for a while, and then follows the northern edge of unincorporated Universal Studios. It continues east, and then bends sharply south as it curves around Griffith Park (at 4,218 acres, one of the largest city parks in the United States). To the north of Griffith Park lies the city of Burbank; to the east is Glendale.

As it flows south, the river runs parallel to another major freeway. You'll see the Silver Lake Reservoir and then Elysian Park, where the Los Angeles Dodgers play Major League Baseball.

Save your bookmarks

Dodger Stadium is a landmark that you may want to return to so you can save the map location (or "extent") as a bookmark.

1) Center your view on Dodger Stadium, more or less as shown in the figure.

2) On the Map tab, click Bookmarks > New Bookmark.

3) In the Create Bookmark dialog box, name the bookmark Dodger Stadium and click OK.

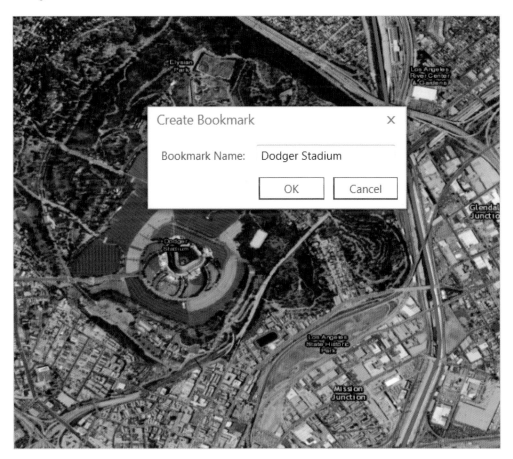

Pan to the city limits

You'll follow the river until it crosses the LA city limits, which marks the boundary of your study area.

1) Pan along the river as it runs south.

This last section of river passes through an industrial landscape and leaves the city at the Redondo Junction train yards.

2) Close the Lesson1a map view (not the entire project) and any open tables.

3) Save your project.

4) Continue to the next lesson or close ArcGIS Pro. Save your changes if prompted.

Results for the book's exercises can be found online on the book's resource web page. For information on how to download and use the Results data, go to esri.com/Understanding-GIS-3.

Exercise 1b: Do exploratory data analysis

In this exercise, you'll add park data and census data (containing demographic and socioeconomic information) to your map. The goal is to pay attention to patterns in the data and thereby build an intuitive sense of likely and unlikely locations for the park. This intuition should give you confidence that the analysis results you get in lesson 6 are plausible. Conversely, if the results contradict your gut feeling, you may be alerted to possible mistakes in the analysis.

Get started

You'll start ArcGIS Pro, if necessary, and continue working with your project document from the last lesson.

1) **Start ArcGIS Pro.**

From the Project pane, you'll make a copy of your Lesson1a map and name it Lesson1b. You can easily copy and paste maps in the Maps section of the Project pane similar to how you copy and paste files on your computer.

2) If necessary, expand the Maps item in the Project pane.

3) Right-click the Lesson1a map and click Copy.

4) Right-click the Maps item (the folder above Lesson1a) and click Paste.

5) Rename the copy of Lesson1a as Lesson1b.

6) Open the lesson1b map by double-clicking the map in the Project pane.

Add a layer of parks

The Parkland feature class can be found in ParkSite > SourceData > ESRI.gdb > Landmark > Parkland.

1) Add the Parkland layer by dragging it to the map.

One of your geographic constraints is that the new park not be located too close to existing parks. You know from your exploration so far that there are a couple of big parks along the river, but now you can look at the whole scenario.

Symbolize the layer

A shade of green is usually the right cartographic choice for parks. ArcGIS Pro may have chosen one by a stroke of luck, but probably not. You'll symbolize the layer after taking a look at its extent.

1) In the Contents pane, right-click the Parkland layer and click Zoom To Layer.

2) In the Contents pane, turn Parkland off and on a few times by clicking its check box. Leave it turned on.

The data extends well beyond Los Angeles, and the basemap shows that those big polygons to the north correspond to mountains. They're probably national forests.

3) In the Contents pane, click the color patch under the Parkland layer name to open the Symbology pane.

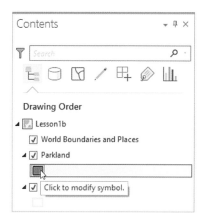

4) Switch from the Gallery tab to the Properties tab and click the Color button to open the color palette. Click Apple Dust, a shade of green.

5) Click the Outline color button to open the color palette again.

6) Click Moss Green.

7) Click Apply in the Symbology pane.

The layer symbology is updated on the map.

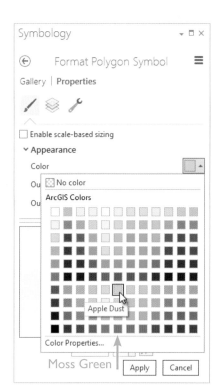

Identify features

Now you can find out what attributes this layer has in the table.

1) On the Map tab, verify that the Explore tool is active by clicking on it.

2) Click on one of the big park features in the mountains to identify it.

Among the attributes are the park's name, its type, and its acreage.

3) **Leave the pop-up window open. On the Map tab, in the Navigate group, click Bookmarks > City of Los Angeles.**

At this scale, you can make out the larger parks within the city.

4) **Try to locate the parks mentioned earlier: the Sepulveda Dam Recreation Area, Griffith Park, and Elysian Park. Click on each one to see its attributes.**

5) **When you're done, leave the Identify window open and zoom to the Dodger Stadium bookmark.**

6) **Identify some of the neighborhood-size parks in the view.**

7) **Close the pop-up window.**

Your list of park requirements doesn't have an upper size limit, but you can expect your candidates to be under 10 acres. Your trip down the river in the previous exercise didn't reveal any great big tracts of open land that weren't already parks.

Label the parks layer

Some of the larger parks are already labeled in the World Boundaries and Places layer, but many of the smaller ones are not.

1) **With the Parkland layer selected in the Contents pane, click the Labeling tab above the map and, in the Layer group, click the Label button on the far left of the ribbon.**

The parks are labeled with their names in simple black text. The information is being taken from the NAME attribute in the Parkland attribute table.

The default labels aren't ideal for this map. The black text disappears into the imagery wherever the label doesn't fit inside the park. In addition, the labels appear on one line, no matter how long they are.

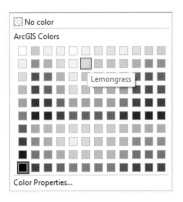

2) On the Labeling ribbon, in the Text Symbol group, change the font to Arial (near the top).

3) Click the Color button. On the color palette, change the text color to Lemongrass.

Light-green text will show up better than black, but to be legible it still needs an outline, or *halo*.

4) Click the launcher button ⌐ in the Text Symbol group to open the Label Class pane.

5) On the Symbol tab, expand the Halo group.

6) Click the selector arrow next to the Halo symbol, click White fill under Polygon symbols, and click Apply (at the bottom of the pane). The default halo color is white, but you're going to use something darker.

7) Click the Color selector under Halo and click Moss Green.

8) Click Apply to see the result on the map.

On the map, the park labels are visible against the basemap.

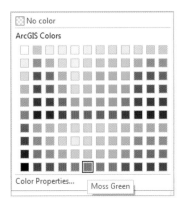

Set a scale range for the labels

The park labels look good at this fairly large scale, where there's room to accommodate them. Now you can see what happens when you zoom out.

1) Zoom to the City of Los Angeles bookmark, located on the Map tab.

At this smaller (zoomed-out) scale, the labels overwhelm the map. You could just turn them off, of course, whenever they seem too crowded. A better solution is to make their visibility depend on the map scale.

2) With the Parkland layer selected, open the Labeling (not Appearance) ribbon.

3) In the Visibility Range group, click in the Out Beyond box and type 40,000 (not 1:40,000). Then press Enter. The park labels turn off instantly because you are zoomed out beyond 1:40,000. Whenever the map scale crosses the 1:40,000 threshold, the park labels will turn off automatically.

4) Zoom to the Dodger Stadium bookmark.

As long as the map scale is larger than 1:40,000 (which means it is zoomed in closer than 1:40,000), the labels show up again.

You've probably noticed that some parks, such as Elysian Park and Echo Park, are labeled in both the World Boundaries and Places and Parkland layers. You can set the scale range of the World Boundaries and Places layer reciprocally to the Parkland labels.

5) **With the top World Boundaries and Places layer selected, open the Appearance ribbon (there should be no Labeling or Data options).**

Scale ranges

As you've just seen, scale ranges can be set both for a layer's labels and for the layer itself. In this case, the distinction was blurred because the Reference layer is composed of labels. But layers that are composed of features, such as parks or rivers, can have their scale ranges set in exactly the same way. When you design maps to be viewed at multiple scales, you normally want layers representing detailed data, such as buildings or utility lines, to be visible only at large (close-up) scales. You may want your map to include multiple copies of a layer, each with a different scale range and unique symbology. For example, one layer might represent trees with a generic point symbol at medium scales. A second layer might represent the same trees with a more detailed and realistic symbol at large scales.

6) In the Visibility Range group, click in the In Beyond box and type 40,001.

Notice that some default options were already set here. These are defaults that are set for this hosted basemap.

By setting this option, the entire top World Boundaries and Places layer turns off when zoomed in at a larger scale than 1:40,001. Most of these labels are actually for places other than parks, but you can accept their loss at large scales.

7) Zoom back and forth across the 1:40,000 scale threshold to test your settings.

At scales of 1:40,000 or larger, the World Boundaries and Places layer appears with a dimmed check box in the Contents pane. This dimmed check box tells you that the layer is turned on but is set not to show at the current map scale.

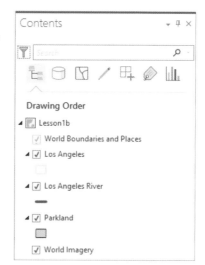

Dim the basemap

Without doing anything further to the labels, you can emphasize them a little more by dimming the imagery basemap. To do that, you'll adjust its transparency.

1) With the bottom World Imagery layer selected, open the Appearance ribbon.

2) From the Effects group, adjust the transparency slider to 10%. You can do so by dragging the slider to 10%, or by typing 10 in the box. On the map, notice that the basemap fades slightly.

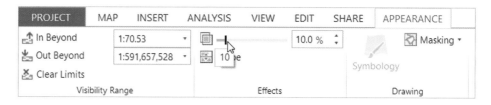

Add a layer of census tracts

The US Census Bureau gathers socioeconomic data about households and aggregates it by various geographic units. One of these units is a *census tract*, which is a relatively small subdivision of a county.

1) Open the Project pane and browse to ParkSite > SourceData > census.

2) Drag the tracts shapefile to the map.

The tracts layer is added to the Contents pane. The tracts cover everything except the labels.

3) In the Contents pane, right-click the tracts layer and click **Zoom To Layer**.

The census tract data covers Los Angeles County, including the islands of Catalina and San Clemente.

4) Zoom to the City of Los Angeles bookmark.

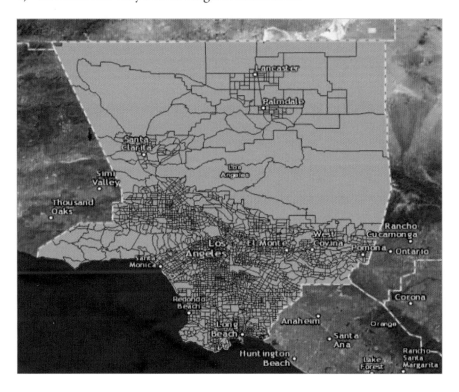

The tracts layer is made up of contiguous polygons that are reminiscent of a jigsaw puzzle. Because each piece of the puzzle represents a different set of living, breathing human beings, it's natural that each tract's attribute values for population, income, age, and so on would be different.

Open the attribute table

Now you can see what attributes the table contains.

1) In the Contents pane, right-click the tracts layer and click Attribute Table.

2) Scroll across the table and look at the field names.

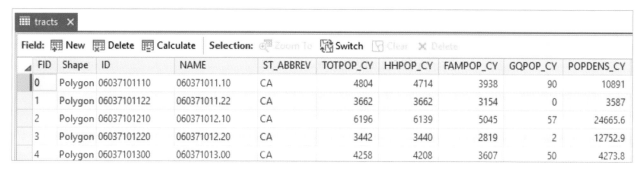

The table has some identification codes, followed by selected population and housing attributes. Some of the field names are fairly easy to interpret, others less so. In lesson 2, you'll see how to get more information about your information by accessing the metadata.

3) Locate the POPDENS_CY field. CY stands for current year, which for this census data is 2015.

This attribute stores population density (people per square mile) for the year 2015. Because one of your criteria is to locate the park in a densely populated area, this is relevant information. The attribute doesn't tell you what value should be a threshold for "high density," but it gives you a way to start making patterns on a map. Close the table when you are finished looking at it.

Symbolize census tracts by population density

Symbolizing a layer by an attribute, also called *thematic mapping*, allows you to see how values are spatially distributed.

1) With the tracts layer selected in the Contents pane, open the Symbology pane by clicking Symbology on the Appearance ribbon.

By default, all features in a layer have a single symbol. That's why all your census tracts are purple, or whatever color they happen to be.

2) In the Symbology drop-down list, click Graduated Colors. The map updates automatically on the basis of some default values.

3) In the Field drop-down list, click POPDENS_CY. On the map, the tracts are symbolized by population density.

4) Change the Color scheme to Yellow to Red.

You can view the names of the schemes by clicking the Show names check box at the bottom of the list. Additional color schemes are available by clicking the Show all check box.

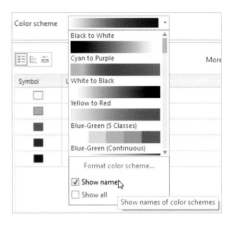

A lot is going on here if you expand the Symbol chart. (The same information is also available in the Contents pane.) The values for the POPDENS_CY attribute, which range from 0 to 96,824.5, are divided into five classes. The starting and ending values for each class are calculated by a "natural breaks" algorithm that separates clumps in the data. That's why the range of values is different from class to class and why classes break at seemingly arbitrary numbers. Each class is associated with a symbol in a color ramp and is displayed on the map.

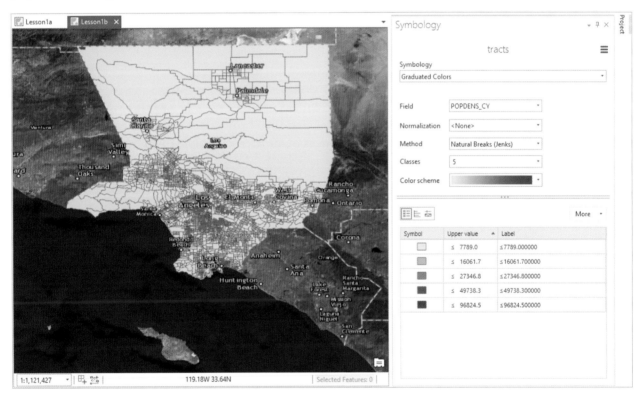

Change the classification

Quantitative symbology is flexible, and you can present data in many ways. Because all you want right now is a general sense of viable areas for your project, and because you're going to look at a couple of variables together, you should keep your presentation simple.

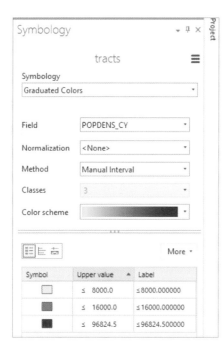

1) Change the Classes drop-down box to 3.

2) Set the classification method to Quantile.

Using three classes, you'll easily be able to see high, medium, and low values. The quantile method guarantees that an equal number of tracts will fall into each class. It should be noted that there are no inherently good or bad ways to classify data—different classifications may be more or less appropriate to the purpose of your map and the background knowledge of your audience.

3) Change the first upper value to 8000 by double clicking the label, typing the value into the Upper value cell, and pressing Enter. The second class break point is selected and editable.

4) Change the second upper value to 16000.

Notice that the classification method has been reset to Manual Interval because you've changed the class breaks. The histogram is updated, too. You no longer have a pure quantile classification, but having your classes break at round numbers makes intuitive sense.

The outlines of the polygons are currently overwhelming the map in some of the areas with the highest densities, making it difficult to interpret the classifications.

5) Click the More button in the Class breaks area of the histogram, and then browse to Symbols > Format all symbols. Any changes made here will apply to all the classifications.

6) If necessary, change to the Properties tab (near the top of the pane). Change the Outline color to No Color.

7) Click Apply at the bottom of the pane, and then click the back arrow at the top of the Symbology pane to return to the classification settings.

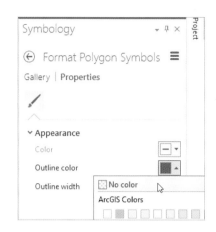

You're removing the outlines because you don't need to see the tract boundaries on the map. For now, you're interested in them as areas, not specifically as tracts.

8) In the Label column, double-click on the first label (< 8000) to make it editable. Type Low and press Enter.

▶ If you start typing in the wrong box (in this case, Upper value), change the classification method to Natural breaks to reset the upper value, and then begin typing in the correct box, Label.

9) Replace the second label with Medium. Press Enter.

10) Change the third label to High. Click outside the edit box to commit the edit.

11) Change the transparency of the tracts layer to 50% (from the Appearance ribbon).

12) In the Contents pane, drag the tracts layer under the Parkland layer.

On the map, you can now see where population is concentrated along the river, and you can see it in relation to existing parks.

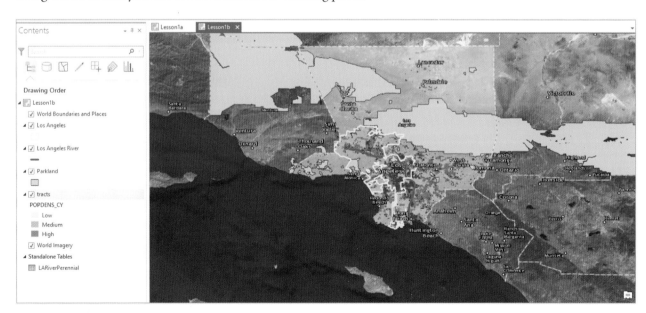

Measure distance from the river to parks

Making a few measurements will improve your ability to estimate distance on the map and give you a better intuitive sense of how close to the river the new park should be.

1) Select the Dodger Stadium bookmark (if necessary, turn off the tracts layer).

2) Pan the map so that a number of parks are in the view. Feel free to zoom in or out.

3) On the Map tab, in the Inquiry group, click the Measure > Measure Distance button to open the Measure Options.

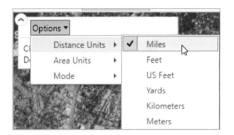

4) Click the Options drop-down arrow. Set the Distance units to Miles and turn off Feet.

On the map, the pointer changes to a ruler with inscribed cross hairs.

5) Move the pointer over a park, such as Cypress Park (northeast of Dodger Stadium on the east side of the river).

6) Click to start a measurement.

7) Move the pointer (you don't have to drag it) to the river. The measurement result is displayed along the line and in the Measure dialog box.

8) Double-click to end the measurement.

9) Click on another park and measure its distance to the river.

The new result replaces the previous one in the Measure dialog box.

10) Measure the distances from a few more parks.

Cypress Park, Elysian Valley Recreation Center Park, and Downey Playground are close to the river. Elyria Canyon Park, a little over three quarters of a mile away (at its nearest edge), stretches the notion of proximity. Bear in mind that these measurements are straight-line distances, not the distance along streets.

11) Switch back to the Explore tool. The Measure tool stays active until another tool is selected.

12) Zoom to the City of Los Angeles bookmark.

Add a layer of census block groups

Another requirement for the new park is that it be located in a lower-income neighborhood. To symbolize an income attribute, you must add another layer to the map. (It's not that income data isn't reported at the census tract level, it's just that your particular tracts layer doesn't happen to include it.)

The same census folder that contains tracts.shp also has a shapefile dataset named block_groups.shp.

1) **Open the Project pane and add the block_groups shapefile to the map from ParkSite > SourceData > census.**

The new layer is added above the other layers. Like the tracts layer, the block_ groups layer covers Los Angeles County. And like the census tracts, the block group polygons resemble a jigsaw puzzle. Block groups are another Census Bureau statistical unit: they're smaller than tracts and nest inside them. We'll talk more about census geography in lesson 2.

2) **Open the attribute table of the block_groups layer and scroll across the attributes.**

Most of these attribute names are cryptic. One of the last fields in the table is called MEDHINC_CY. For now, take it on faith that this acronym stands for median household income.

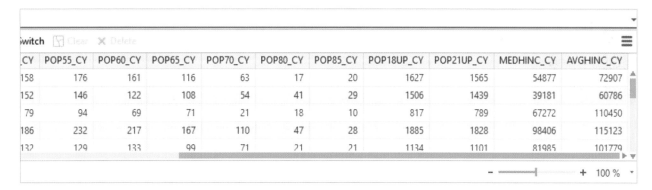

3) **Close the attribute table.**

Symbolize census block groups by median household income

If you symbolize the block_groups layer with graduated colors, you won't be able to evaluate income and population density at the same time. Instead, you'll represent each block group's median household income as a point drawn inside the block group polygon. The point sizes will be graduated according to the income value.

1) Using the Appearance tab, open the Symbology pane for the block_groups layer.

2) In the Symbology drop-down list, click Graduated Symbols.

3) In the Field drop-down list, click MEDHINC_CY.

4) Click the Classes drop-down arrow and click 3.

5) Replace the first value with 50000 and the middle value with 100000 and press Enter.

6) Click the Template symbol (underneath the Minimum size drop-down box) to open the Format Point Symbol pane.

7) Under Properties, click the Color button and change its color to Tourmaline Green.

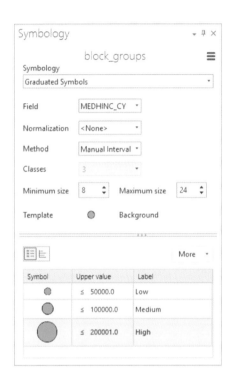

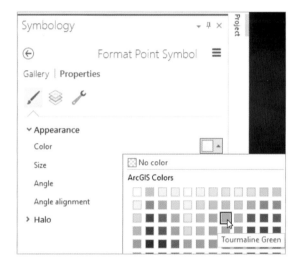

8) Click Apply and then the back button at the top of the pane. This symbology sets the color and shape of the symbols that will be used to represent income values.

9) Click Background (on the right of Template) to open the Format Point Symbol pane again.

10) Change Color to No Color, if necessary, and then change the Outline color to No Color.

11) Click Apply and then the back button at the top of the pane.

This symbology makes the block group polygons themselves invisible. All you'll see are the income dots spread around the map.

12) Change the Minimum size to 8.

13) Change the Maximum size to 24.

14) In the Label column, double-click on the first label (< 50000) to make it editable. Type Low and press Enter.

15) Replace the second label with Medium. Press Enter.

16) Change the third label to High. Press Enter.

At the present scale, the symbols overwhelm the map.

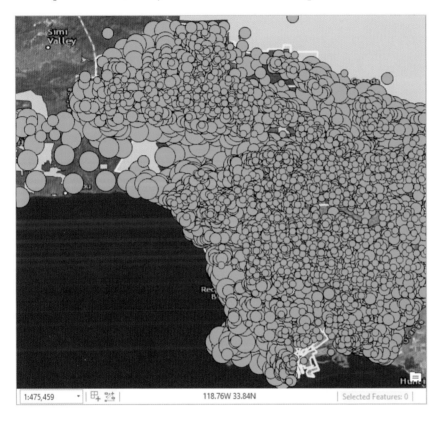

Set a scale range for the block_ groups layer

Earlier, you set a maximum scale value for the World Boundaries and Places layer. Here, you'll set a minimum value for the block_ groups layer.

1) With the block_groups layer selected in the Contents pane, open the Appearance ribbon.

2) Click in the Out Beyond box and type 100000. Press Enter.

The symbols disappear from the map. In the Contents pane, the layer's dimmed check box indicates that the layer is turned on but is not visible at the current map scale.

3) Zoom to the Dodger Stadium bookmark.

4) Click the Fixed Zoom Out button a couple of times until some of the Medium and High symbols begin to appear.

5) Turn the tracts layer back on. if necessary.

Now you can start to get a general sense of household income and population density along the river, and look at these variables in relation to park locations.

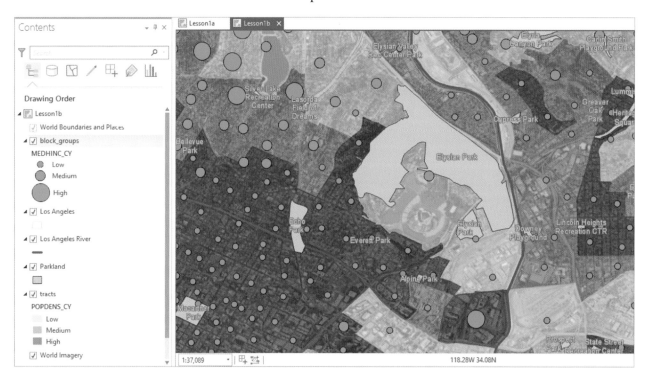

Search for likely park areas

Clearly, you're taking an incomplete initial look at a complex problem. You haven't considered all the requirements (for example, the presence of children). You're not making any exact measurements of distance. Your data classifications are casual: you don't yet have a good reason to say what values should count as high population density or low median household income in the context of your project. Nevertheless, you can form some meaningful impressions. You won't be able to say of an area that a park should definitely go there, but you might be able to identify likely and unlikely areas. Later, it will be interesting to see how well these impressions are borne out by analysis.

1) In the Map Scale box at the bottom of the map pane, highlight the current value. Type **40000** and press Enter.

2) Pan south to where the river crosses the city boundary.

3) In the Contents pane, drag the Los Angeles River layer to the position just below the World Boundaries and Places layer (it may already be there).

You'll follow the river to its source, marking good areas for a park along the way. A really good area would have these properties:

- high population density (dark red),
- low median household income (small green dot),
- no existing park nearby, and
- close to the river.

4) Pan slowly north.

Less than a mile north of the city limits, on the east side of the river, are a couple of tracts—one dark red and one medium red—with small green dots. They're pretty close to the river, and even though there are some parks in the general vicinity, it's probably worth marking the area as a potential park.

Exercise 1b: Do exploratory data analysis

To mark the area, you'll use a feature of ArcGIS Pro called Map Notes. Map Notes are layer templates that create point, line, and polygon feature classes in your Home database with predefined symbols for feature creation and editing.

5) On the Insert tab, in the Layer Templates group, click the Bright Map Notes button.

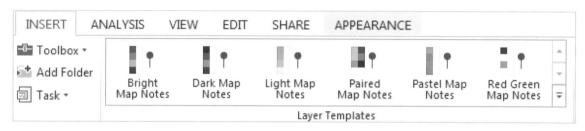

Clicking the button adds a new group layer to your Contents pane named Bright Map Notes.

6) Expand the Bright Map Notes layer to see that it contains point, line, and polygon layers.

7) Expand the three layers to see their predefined symbols.

8) Collapse the Bright Map Notes group by clicking the little arrow on the left of the check box.

9) In the Project pane, browse and expand your home database in the Databases group. Notice that three new feature classes have been added to the geodatabase. These classes are the data sources for the Map Notes layers that were added to the map.

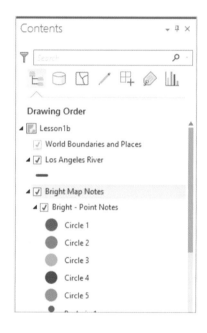

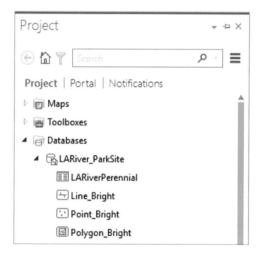

Create a new feature

You will now create a new feature in the feature class.

1) Go to the Edit tab.

2) In the Features group, click the Create button.

This button opens the Create Features pane on the right of the map. This pane allows you to draw new features in any of the layers marked as editable in the map. By default, all layers are editable so you will change that so that you don't accidentally add new rivers, parks, census tracts, and so on.

3) In the Contents pane, click the List By Editing button above the layer list.

This list now displays all the layers with a check box next to each one to define whether they are editable or not.

4) Click to clear all but the Bright Map Notes layers (you will probably have to expand the group layer to confirm that they are checked).

Notice that as you click to clear the other layers, the Create Features pane on the right of the map changes to show only those features that are editable.

5) Switch back to the List by Drawing Order view by clicking the button at the top of the Contents pane.

6) Add a purple star point near the areas of interest by clicking the Star 4 icon from the Create Feature pane on the right side of the application and then clicking the spot on the map.

▶ To stop adding stars, click the Explore button. To delete unwanted stars, click the Undo button.

Continue searching for park areas

You'll keep following the river and looking for likely places for a park.

1) Click the Explore button on the Map tab and pan north.

A little farther north, again on the east side of the river, is a medium-density, lower-income area with just one park (Pecan Playground) in the general vicinity. You'd prefer a high-density neighborhood, but since you don't yet have specific criteria for your analysis, you want to be inclusive rather than exclusive in your assessments. It's up to you whether to mark this area or not—all that really matters is that you start to gain a sense of the study area.

2) Add another purple star in this area by going back to the Edit tab and clicking the Star 4 icon.

The area opposite Dodger Stadium, on the east side of the river, is medium density, lower income, and park poor, which makes it rich with possibility for a new park site.

3) Add another purple star in this area.

4) Mark any other areas that seem promising to you.

As you navigate around Griffith Park, remember that the yellow line marks the city limits. The areas north and east of the park aren't part of Los Angeles and shouldn't be considered. Once you get around this park, you may find fewer likely areas. But remember, there's no right or wrong answer for this exercise.

5) When you're finished marking promising locations, zoom to the layer Bright – Point Notes.

6) Save your edits by clicking the Save button in the Manage Edits group.

7) When prompted to save all Edits, click Yes.

8) Close the Create Features pane.

9) Close the Lesson1b map and any open tables.

10) Save your project.

11) **Continue to the next lesson or close ArcGIS Pro. Save your changes if prompted.**

Now that you've explored the study area around the Los Angeles River, you will begin to list the data requirements for the project and begin previewing the data that meets your project needs. This data can also help you reframe your problem and park criteria in more detail by replacing general guidelines with hard numbers and thresholds.

Lesson 2 Preview the data

YOU NEED THE RIGHT DATA TO solve the problem of where to locate a park. You explored the study area in lesson 1. Now you can proceed more systematically. What data do you have? How useful is it? Is there data that you need but don't have? Has the problem been stated clearly enough for you to know what data you need?

Acquiring, evaluating, and organizing data is a big part of an analysis project. This book doesn't fully re-create the complexity of the real world because we've provided all the basic data required. On the other hand, much of the data isn't project-ready, and that need for further preparation reflects the real world of GIS.

The first thing you'll do in this lesson is draw up a planning document to help keep your tasks in focus. You'll use this document to list the guidelines for the new park and translate them into specific needs for spatial and attribute data.

After you itemize your data requirements in general terms (park data, river data, and so on), you'll take stock of your source data and investigate its spatial and attribute properties. You'll also familiarize yourself with metadata, which is the data you have about your data. Before you decide to use a particular dataset, you may want to know things such as who made the data, when, and to what standard of accuracy.

Once you have a better working knowledge of your data, you'll reframe the problem statement. GIS is a quantitative technology: you can't analyze a problem until it's been stated in measurable terms. Wherever you find the city council's guidelines to be vague, you'll replace them with hard numbers.

Exercise 2a: List the data requirements

You must relate the guidelines for the new park to data requirements for the project.

Lesson Two road map

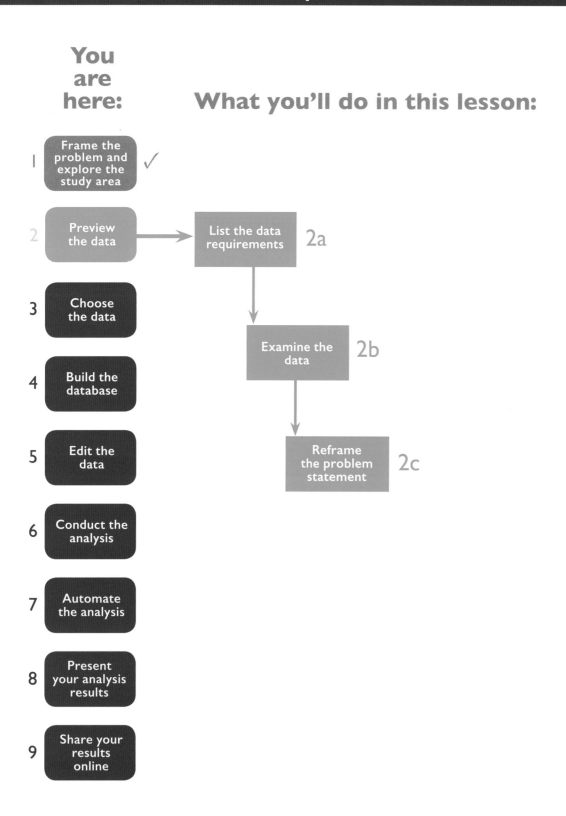

Open the data requirements table

We've made a table in advance to help you keep track of your requirements. It's an informal document, but it will still be helpful. You can refer to it as the data requirements table.

1) Open Windows Explorer and navigate to C:\UGIS\ParkSite\MapsAndMore.

2) Double-click the file DataRequirementsTable.doc to open it in Microsoft® Word.

▶ If you don't have Microsoft Word, open the RTF version of the document in another application, or print the PDF version and fill it out with a pencil.

#	REQUIREMENT	DEFINED AS	SPATIAL DATA	ATTRIBUTE DATA	DATASET	PREPARATION
1	land parcel		parcels			
2	vacant					
3						
4						
5						
6						
7						
8						
9						
10						
11						
12						
13						
14						
15						

Pieces of land, legally surveyed, described, and recorded with a county or other administrative body, are called *parcels*.

List the requirements

In this section, you'll review the city council's guidelines (refer to "Park guidelines" in lesson 1, under "Frame the problem") and describe in a general way the data needed to satisfy them. We'll get to the specifics of choosing datasets in lesson 3.

The first guideline was to find a vacant piece of land at least one-quarter acre in size. You can break this down into three requirements:

- Land parcel
- Vacancy
- Size

The requirement for a land parcel is already listed in the table. You need spatial data representing parcels so that you can see candidate sites on the map.

The second requirement is vacancy, which is a characteristic, or attribute, of a parcel. In a GIS dataset, vacancy is often listed with other descriptions of land use (commercial, residential, industrial, and so on). In general terms then, you're looking for a land-use attribute.

1) **In row 2 of the table, under Attribute Data, type (or write) land use.**

The third requirement is that the park be one-quarter acre or larger. Like vacancy, acreage is an attribute, although it is one that can be calculated by the software. Because ArcGIS Pro can convert one unit of area to another, you don't even have to start with acres—any measurement of parcel size will suffice.

2) **In row 3, under Requirement, type a quarter acre or more. Under Attribute Data, enter area.**

The second guideline under "Park guidelines" is that the park be within the Los Angeles city limits. This sounds like spatial data, and you'll treat it that way for now. (It could be an attribute, too, because a field in a table might store the name of the city in which each parcel is recorded.)

3) **Fill in row 4 as you think it should look, and then check the figure.**

#	REQUIREMENT	DEFINED AS	SPATIAL DATA	ATTRIBUTE DATA
1	land parcel		parcels	
2	vacant			land use
3	a quarter acre or more			area
4	within city limits		cities	
5				

The third guideline is that the park be as close as possible to the Los Angeles River.

4) **In row 5, for the requirement, put near LA River. Under Spatial Data, put rivers.**

Using spatial datasets of parcels and rivers, you can measure the distance from any given parcel to the river.

The fourth guideline is to locate the park not in the vicinity of another park, or away from existing parks.

5) **Fill out row 6 as you think it should look.**

The fifth guideline also needs to be broken down. You need a neighborhood (spatial data) that has the following:
- high population density (attribute data) and
- lots of children (attribute data).

Neighborhoods tend not to have formal boundaries, so you're probably not going to find them as such in a spatial dataset. As a proxy, or substitute, you'll use a set of small, standardized areas defined by the US Census Bureau: either the tracts or block groups you looked at in lesson 1.

6) In row 7, enter in a neighborhood as the requirement. Enter census unit for the spatial data.

7) In row 8, enter densely populated for the requirement and population density for the attribute data.

8) In row 9, enter lots of kids for the requirement. For the attribute data, enter age.

The sixth guideline is that the park be in a lower-income neighborhood. You don't need to repeat the spatial requirement for a neighborhood from step 6.

9) In row 10, enter lower income for the requirement and income for the attribute data.

The last park guideline is to serve as many people as possible. For this guideline, you need a population attribute.

10) In row 11, enter serving the most people as the requirement and population as the attribute data.

#	REQUIREMENT	DEFINED AS	SPATIAL DATA	ATTRIBUTE DATA	DATASET	PREPARATION
1	land parcel		parcels			
2	vacant			land use		
3	a quarter acre or more			area		
4	within city limits		cities			
5	near LA River		rivers			
6	away from other parks		parks			
7	in a neighborhood		census unit			
8	densely populated			population density		
9	lots of kids			age		
10	lower income			income		
11	serving the most people			population		
12						
13						
14						
15						

Eventually, you'll want to make a map of potential sites, and you may need some data for cartographic purposes. For example, political boundaries and roads put maps in a familiar context. Physical relief creates texture, and imagery provides realistic detail.

11) In rows 12 to 15, enter final map for the requirement. Under Spatial Data, list the examples just mentioned in step 10.

#	REQUIREMENT	DEFINED AS	SPATIAL DATA	ATTRIBUTE DATA	DATASET	PREPARATION
1	land parcel		parcels			
2	vacant			land use		
3	a quarter acre or more			area		
4	within city limits		cities			
5	near LA River		rivers			
6	away from other parks		parks			
7	in a neighborhood		census unit			
8	densely populated			population density		
9	lots of kids			age		
10	lower income			income		
11	serving the most people			population		
12	final map		political boundaries			
13	final map		roads			
14	final map		relief			
15	final map		imagery			

12) Save and minimize the table. You'll continue to use it in the next exercise.

Exercise 2b: Examine the data

Now you can see what data you actually have on hand. To do this, you'll work in the Project pane. In lesson 1, you used the Project pane to manage your maps and data folders. The Project pane is great for going back and forth between your map and your data (which is what you do most of the time).

Get started

1) Start ArcGIS Pro and open your LARiver_ParkSite project. You'll start ArcGIS Pro, if necessary, and continue working with your project document from exercise 1b.

2) Display the Project pane. The Project pane is generally docked on the right of the map (sometimes hidden as a tab).

▶ If you close the Project pane or can't find it, you can always open it by clicking the Project button on the View tab.

Insert a new map from the Project pane

Recall that you can insert a map from the ribbon, but this time you'll add it from the Project pane.

1) In the Project pane, expand the Maps folder . Note that the map(s) you've created in previous lessons are listed here.

2) Right-click Maps in the Project pane and click New Map in the context menu. A new map is added to the list and open to the map view.

3) Right-click the new map item and click Rename. Name the new map Lesson2.

4) In the Project pane, expand Folders by clicking the arrow on the left of it (or double-click Folders). Here you should see the project folder that was created when you started the project (LARiver_ParkSite) as well as the ParkSite folder that you connected to in lesson 1.

Survey the SourceData folder

1) Expand the ParkSite folder.

2) Expand the SourceData folder.

3) Expand everything you can under SourceData.

It's a long list of items. You may have to scroll down or maximize the application to see everything. Each item is a piece of geographic data or a data container. The icons signify the type of data, as illustrated in the sidebar "Representing the real world as data" on the next page.

Under the SourceData folder are three folders and a geodatabase .

- The census folder contains three feature classes of census data in shapefile (.shp) format.
- The City of LA folder contains three shapefiles and a stand-alone table in dBASE (.dbf) format.
- The ParkData folder contains a shapefile.
- The geodatabase contains 10 feature classes in geodatabase format. The feature classes are thematically organized in containers called *feature datasets* .

In the next several sections, you'll preview a lot of this data to make sure you have the features and attributes you listed in the data requirements table.

Exercise 2b: Examine the data   65

Representing the real world as data

How would you create an information system to organize and manage the huge variety of geographic stuff in the world? One approach is to think of all that stuff in terms of discrete objects.

The discrete-object view of the world

If you conceive of geography in terms of objects, you can sort these objects by similarities. Shape is a fundamental sorting principle: every object can be drawn—in two dimensions—as either a point, a line, or a polygon. Theme, or type, is another principle: every object can be classified as a school, a road, a park, or something else.

Applying these sorting principles of shape and theme, you can come up with collections of things you would recognize on a map: schools represented as points, roads represented as lines, parks represented as polygons, and so on.

Points, lines, and polygons representing geographic objects on a map.

Each object in a collection has a unique location, specified by a pair of spatial coordinates (for points) or a list of coordinate pairs (for lines and polygons).

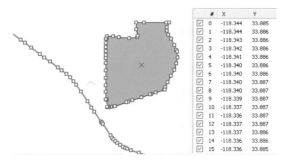

A polygon (with an *x* indicating its centroid) and the list of coordinate pairs that defines its location.

Besides a unique location, every object has a set of facts that pertain to it: a name, a description, or whatever bits of information have been gathered about it. These facts are the object's attributes.

FID	Shape *	PARK NAME	CATEGORY	ADDRESS	TELEPHONE	HOURS
0	Polygon	Los Angeles	State Historic Park	1245 N. Spring Street Los	213-620-6152	8 AM - sunset
1	Polygon	Rio de Los Angeles	State Recreation Area	1900 San Fernando Road	213-620-6152	9 AM - 10:30 PM
2	Polygon	Vista Hermosa Park	Local Park or Recreation Area	100 N. Toluca Street Los	213-250-3578	8 AM - sunset

In ArcGIS, a collection of such objects—with a common shape, common theme, and common attributes—is called a *feature class*. An individual object in the collection is a *feature*. The feature class is the basic storage unit for GIS data created according to the discrete-object view of the world, commonly called the *vector data model*.

Feature classes can be stored in various file formats, notably the geodatabase and the shapefile. The geodatabase format is newer and more highly developed.

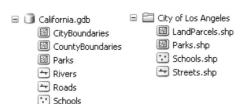

Feature classes in geodatabase format (*left*) and shapefile format (*right*). Different icons indicate polygon, line, and point data.

The continuous-surface view of the world

Although it's a powerful model, the discrete-object view is not an intuitive way to think of certain kinds of geographic information, such as elevation or temperature, that don't have shapes or boundaries and that cover the world everywhere. It's quite possible to represent these phenomena as features (for example, contour lines represent elevation on topographic maps), but a more natural way to think of them is in terms of continuous expanses, or surfaces.

The most common way to model a geographic surface is with a matrix of square cells, or pixels. Each cell represents a unit of area, such as a square meter, and stores a single piece of geographic information—typically, a measured or estimated value—at that location.

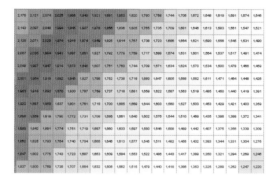

In this example, each cell represents 30 square meters of ground and stores a single representative elevation value.

This way of modeling surfaces is called the *raster data model*. It's commonly used for elevation and its derivations (slope, aspect); for temperature, precipitation, and land cover; for statistical data, such as densities and means; and especially for imagery.

The raster dataset is the basic storage unit for GIS data created according to the continuous-surface view of the world. Raster datasets can be stored in geodatabases or in various standard image file formats, such as TIFF and JPEG.

Raster datasets in geodatabase format (*left*) and .tif format (*right*).

Feature classes and raster datasets are complementary. In many maps, raster datasets are used for background display, whereas feature classes are used for foreground display and analysis.

Exercise 2b: Examine the data 67

Preview parcels

First, you'll preview the parcels data, which comes from the County of Los Angeles.

1) **Switch to the Imagery with Labels basemap using the Basemap button on the Map tab.**

2) **In the Project pane, under the City of LA folder, click Parcels.shp to highlight it.**

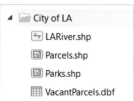

The icon 🔲 signifies a polygon feature class in shapefile format.

3) **Right-click Parcels.shp and click View Metadata.**

A new Project tab is displayed and shows metadata, or data documentation. What you see here is the Details, an overview of the dataset. Complete metadata includes a description about the data, when and how it was created, attribute information, and so on.

Acquiring data

In any GIS project, acquiring good data is a big part of your job. ArcGIS℠ Online provides datasets representing many types of geography. You can access this data from a web browser (www.arcgis.com) or directly from ArcGIS Pro by switching from the Project tab on the Project pane to the Portal tab. You can search for particular types of data by keyword; you can also search for ArcGIS Online groups, such as Living Atlas, that curate a variety of authoritative content. For spatial data, you can use ArcGIS℠ Open Data open-source software. Spatial data is also widely available from government agencies, educational institutions, and commercial vendors. All these sources may supplement data collected and managed by your own organization.

City of Los Angeles Parcels

Type Shapefile

Tags Los Angeles | LA | County of LA | landbase | parcels | landbase parcels | right of way | subdivision | lot | deed | tract | property | cadastral | survey | streets | owner | planning

Summary
Property boundaries for all properties in the City of Los Angeles.

Description
Property boundaries for all properties in the City of Los Angeles. A subset of parcels within the City of Los Angeles was created from the Los Angeles County dataset.

Credits
Los Angeles County Open Data By lahub_admin
Source http://public.gis.lacounty.gov/public/rest/services/LACounty_Cache/LACounty_Parcel/

Use limitations
License This work is free of known copyright restrictions. The City of Los Angeles is neither responsible nor liable for any viruses or other contamination of your system nor for any delays, inaccuracies, errors or omissions arising out of your use of the Site or with respect to the material contained on the Site, including without limitation, any material posted on the Site. This site and all materials contained on it are distributed and transmitted "as is" without warranties of any kind, either express or implied, including without limitation, warranties of title or implied warranties of merchantability or fitness for a particular purpose. The City of Los Angeles is not responsible for any special, indirect, incidental or consequential damages that may arise from the use of, or the inability to use, the site and/or the materials contained on the site whether the materials contained on the site are provided by the City of Los Angeles, or a third party.

Extent
West -118.953532857   East -117.644108646
North 34.8235365168   South 32.7922911738

Scale Range

Details | Preview

4) Scroll down through the Details. From the top, you see:
   - The dataset name (City of Los Angeles Parcels) and file type (shapefile)
   - A thumbnail image
   - Tags that make the data searchable
   - A summary of the intended use of the data
   - A description of the data content
   - Credits attributing where the data came from
   - Use limitations related to the license agreement
   - Extent of the dataset in latitude and longitude
   - Scale range of the dataset from maximum to minimum

5) Switch from the Details view to the Preview view in the lower-left corner of the Parcels metadata. This view allows a visual inspection of the data. The view can be panned and zoomed.

6) Close the Project View tab (currently displaying the metadata).

7) Add the Parcels shapefile to the Lesson2 map. The map will zoom to the extent of the shapefile.

8) Zoom in until you can easily distinguish features.

▶ The Parcels dataset is large, so the response may be a bit slow. Shapefiles can also be considerably slower than other formats.

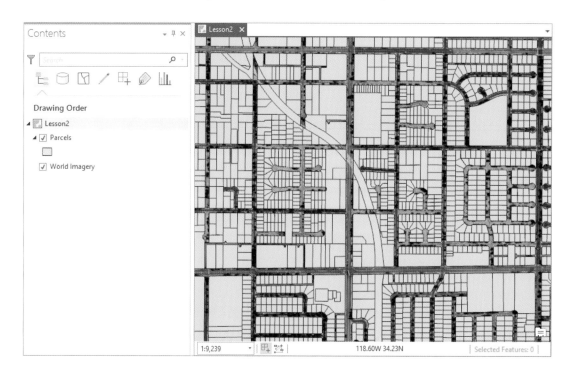

This is the spatial data you need, showing individual parcel boundaries.

9) With the Explore tool enabled on the Map tab, click any parcel to identify it. The results will be displayed in a pop-up window showing all the attributes of the feature.

Depending on exactly where you click, you may identify one or more features. The number of identified features will be displayed on the lower left of the pop-up window. You can use the arrow buttons at the bottom to move through the features.

The vacancy attribute you need isn't here, but you do have an area attribute in unspecified units. Once you find out what the units are (which you'll do in lesson 3), you can convert them to acres. The figure provides some visual context for the size of a one-quarter-acre parcel compared with a five-acre park.

10) Close the pop-up window.

Preview the table of vacant parcels

Since you don't have a vacancy attribute in the parcels shapefile, you'll look for it elsewhere.

1) In the Project pane, under the City of LA folder, right-click VacantParcels.dbf and click Add to Current Map.

This stand-alone table has information about parcels, but no polygons or any other association with spatial features. Stand-alone tables are added to the Contents pane in a special group labeled Standalone Tables.

2) Open the table by right-clicking it in the Contents pane and clicking Open. Each of the 29,461 records represents a vacant parcel within the city of Los Angeles.

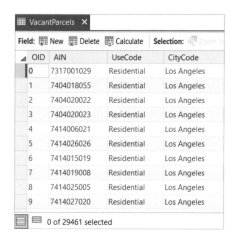

3) Read the column headings.

OID (object identifier) is a sequential number created and managed by ArcGIS automatically. AIN (assessor identification number) is a user-managed identifier. The UseCode field identifies the parcel use. The CityCode field identifies the city in which the parcel is located.

4) Close the VacantParcels attribute table.

Preview cities

Row 4 of the data requirements table lists cities as needed spatial data. You have this data: you used it in lesson 1, when you made a definition query on Los Angeles.

1) In the Contents pane, turn off the Parcels layer.

2) In the Project pane, under Folders > SourceData > ESRI.gdb, under Boundary, add the City_ply feature class to the Lesson2 map.

3) Right-click the City_ply layer in the Contents pane and click Zoom to Layer. You zoom out to a view of all the features in the feature class, which is the entire United States.

4) Turn off the City_ply layer.

5) Add City_pt from the Project pane (Folders > SourceData > ESRI.gdb > Boundary).

6) Zoom to the City_pt layer.

The main difference between the two feature classes is that City_pt represents cities as points rather than polygons. (Hence the names City_pt and City_ply.) There's also a difference in spatial extent, because City_pt includes a feature for Attu Station in Alaska, a place which is so far "west" that it is actually in the Eastern Hemisphere

7) Zoom in on the data to see individual features.

Why would the same features, such as these cities, be represented with two different types of shapes (points here and polygons in City_ ply)? It's because each shape is appropriate for maps of different scale. For a national map, you would show cities as points. For a local map, you might show them as polygons. For your requirement, which is to make sure that potential park sites lie inside the city limits, you need features that have boundaries—and hence polygons rather than points.

8) Turn off the City_pt layer.

Preview the LA River

In row 5 of the data requirements table, you have rivers as needed spatial data. In lesson 1, you added the River feature class from the ESRI geodatabase. You also have another feature class to look at: LARiver.shp in the City of LA folder.

1) Add the LARiver shapefile to the Lesson2 map.

2) Zoom to the layer (right-click LARiver and click Zoom To Layer).

3) Open the attribute table and scroll to the bottom of the table.

In fact, the Los Angeles River is the only river in the feature class, but it's composed of 265 separate features (the FID, or feature ID, of the first record is 0). Why so many? As noted in lesson 1, the answer has to do with attributes.

4) Review the table to see the attributes, and then scroll up and notice the different values in the Capacity, Discharge, and Protection fields. Many of the rows at the top even have empty attribute values.

For your purposes, it doesn't really matter what these attributes mean. The point is, if the river was represented as a single feature, it would also have just one row in the attribute table. That would mean that only a single value could be stored for each attribute—fine for the river name (which doesn't change), but a problem for anything you might want to measure or describe at different locations along the river: flow, depth, water chemistry, navigability, or anything else. The creators of this data wanted to gather facts about the river at different places. To do that, they had to define the river as a spatially connected series of individual features.

You don't have that need. All you want is the spatial data. If you end up using this feature class in your analysis (rather than the River feature class in ESRI.gdb), you'll probably combine the 265 features into one.

5) Turn off LARiver and close its attribute table.

Preview parks

In row 6 of the data requirements table, you need spatial data representing parks. You already know you have parks data: in lesson 1, you symbolized and labeled the Parkland feature class. There's also a shapefile named Parks in the City of LA folder.

1) Add the Parks layer to the Lesson2 map.

Software-managed attributes

Every shapefile feature class has FID and Shape attributes that are created and managed by the software. The FID (feature identification) attribute stores a unique number for each feature. The Shape attribute stores the geometry type. Behind the scenes, it also links each feature to coordinates that define its spatial location. Measurement attributes, such as length and area, can be calculated for shapefiles, but if the values change—because of a spatial edit, for example—the software doesn't update them automatically.

A geodatabase feature class has up to four software-managed attributes. Like a shapefile, it has a feature identifier (called OBJECTID instead of FID) and a Shape attribute. The Shape_Length attribute stores the length of line and polygon features. This attribute doesn't exist for point features. The Shape_Area attribute stores the internal area of polygon features. It doesn't exist for point or line features. Shape_Length and Shape_Area are automatically kept up to date by the software.

2) Zoom to the layer.

3) Open the attribute table.

The table preview shows that there are 337 records (features) and just a few attributes.

4) Turn off the Parks layer and close its attribute table.

5) In the Project pane under the ParkData folder, add NewParks.shp to the Lesson2 map.

6) Zoom to the NewParks.shp layer.

7) Open the attribute table.

This shapefile has just two features. One is Los Angeles State Historic Park, and the other is Rio de Los Angeles State Recreation Area. Note the absence of length and area attributes. You can create them if you want—for example, the Parks shapefile has them—but they don't exist by default because this is a shapefile format.

8) View the metadata for NewParks.shp (right-click it in the Project pane) and read the summary.

The data represents two newly developed parks. In lesson 3, when you choose a parks feature class for the analysis, you'll have to make sure that it incorporates these two parks.

9) Turn off the NewParks.shp layer and close its attribute table.

We are also aware of a third park, Vista Hermosa Park, that has been completed but is not in NewParks.shp. Located just north of downtown, it's one more park to keep track of.

10) Click the Locate tool 🔍 located on the Map tab, in the Inquiry group, and search for Vista Hermosa Park in the Locate pane.

The park location will be marked on the map as an Ⓐ.

11) Zoom in for a closer look of the park.

12) Bookmark it as Vista Hermosa Park.

13) Close the Locate pane.

Preview census units

In row 7 of the data requirements table, you decided to use census units as a proxy for neighborhoods.

1) In the Project pane, under the census folder, add tracts.shp to the map and zoom to the layer.

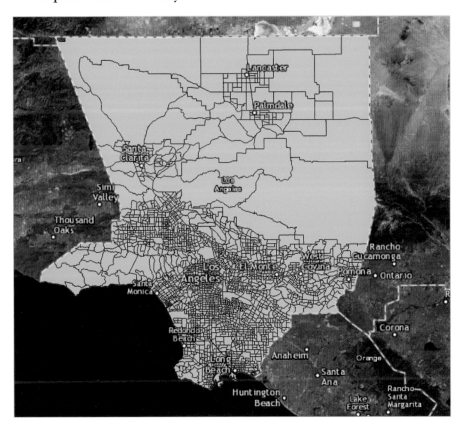

The data covers Los Angeles County. The tracts to the north are much bigger than the ones to the south. That's because census tracts are designed to have a fairly consistent population range, and the northern part of the county, with mountains and desert, is less densely populated.

2) Zoom in somewhere on the southern part of the city.

3) Add block_groups.shp to the map.

4) Add block_centroids.shp to the map. If necessary, zoom in further to see individual points.

Tracts are subdivided into block groups, and block groups are subdivided into blocks. A block centroid is a block represented spatially as a point rather than a polygon. (That's not so strange—you've seen the same thing with cities.) You can learn more about census units in the sidebar "Fundamentals of US Census geography" later in the lesson.

Either block groups or tracts will satisfy your spatial data requirement for a neighborhood. Because of their point geometry, block centroids won't.

Thus far, you've confirmed that you have the spatial and attribute data listed in rows 1–7 of the data requirements table. In three cases (rivers, parks, and census units), you'll have to choose between feature classes. You'll tackle that problem in lesson 3. You still have more data requirements to consider and more data to preview. You must also review your requirements for specificity. You'll do this in the next exercise.

5) Save your ArcGIS Pro project.

Exercise 2c: Reframe the problem statement

Some of the city council's park guidelines are specific and measurable:

- On a vacant land parcel one-quarter acre or larger
- Within the LA city limits

Others are vague:

- As close as possible to the LA River (Is there a maximum allowed distance from the river? If so, what is it?)
- Not in the vicinity of an existing park (How close is "in the vicinity"?)
- In a densely populated neighborhood with lots of children (How densely populated? How many children?)
- In a lower-income neighborhood (How is "lower income" defined?)
- Serving as many people as possible (How big an area does a park "serve"?)

You can't do the analysis until you eliminate the vagueness.

Define proximity to the LA River

Unless you set a maximum distance limit, every vacant parcel in Los Angeles becomes a potential park candidate. That's absurd and could waste a lot of data-processing time. You'll set one-half of a mile as an arbitrary outer limit. That stretches the idea of proximity somewhat, but it's just a cutoff point. Hopefully, you'll find some good locations that are closer than that.

1) From Windows Explorer, navigate to C:\UGIS\ParkSite\MapsAndMore and open the DataRequirementsTable.doc (or the .rtf file if you don't have Microsoft Word).

2) In row 5 of the data requirements table, in the Defined As column, enter <= 0.5 miles.

The symbol <= means "less than or equal to."

#	REQUIREMENT	DEFINED AS	SPATIAL DATA	ATTRIBUTE DATA	DATASET	PREPARATION
1	land parcel		parcels			
2	vacant			land use		
3	a quarter acre or more			area		
4	within city limits		cities			
5	near LA River	<= 0.5 miles	rivers			
6	away from other parks		parks			
7	in a neighborhood		census unit			
8	densely populated			population density		
9	lots of kids			age		
10	lower income			income		
11	serving the most people			population		
12	final map		political boundaries			
13	final map		roads			
14	final map		relief			
15	final map		imagery			

Define "away from other parks"

What minimum distance should a candidate site have to be from existing parks? In open-space planning, 0.25 miles is often used to define a convenient walking distance. (That's typically about a five-minute walk.) Following that standard, you can say that a site is not in the vicinity of an existing park if the nearest park is at least 0.25 miles away.

1) In row 6 of the data requirements table, in the Defined As column, enter >= 0.25 miles.

This measure is a simplification because it's based on straight-line distance.

Fundamentals of US Census geography

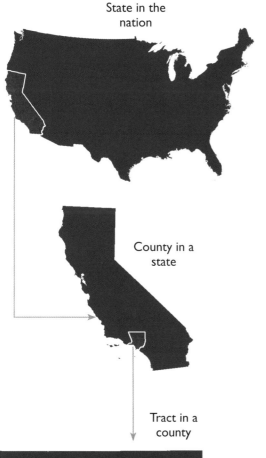

The US Census Bureau reports data by various geographic units. The top-to-bottom relationship shown here represents containment: the nation contains states, states contain counties, counties contain tracts, tracts contain block groups, and block groups contain blocks. Many other nonnesting reporting units are not shown.

Census tracts are relatively small subdivisions of a county. They typically have between 1,000 and 8,000 inhabitants and vary in size. They are designed to be fairly homogeneous with respect to demographic and economic conditions.

A block group is a cluster of blocks within a tract. A block group typically has between 600 and 3,000 inhabitants.

A census block (commonly an ordinary city block) is an area bounded by visible features, such as streets or railroad tracks, or by invisible boundaries, such as city limits. A block centroid is a census block represented as a point rather than a polygon. A centroid is located in the geographic center of the block it represents and has the attributes of that block.

The Census Bureau conducts a new census every 10 years. The latest one was conducted in 2010. Professional demographers estimate values for the intervening years.

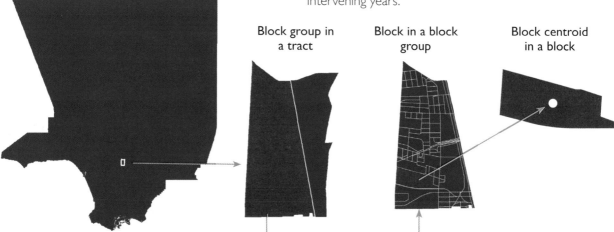

Define a "densely populated" neighborhood

As you make the rest of the requirements concrete, you'll also make sure that you have the appropriate data.

1) Open the attribute table of the tracts.shp layer and scroll across its attributes.

T_ABBREV	TOTPOP_CY	HHPOP_CY	FAMPOP_CY	GQPOP_CY	POPDENS_CY	TOTHH_CY	POP0_CY	POP5_CY	POF
	4804	4714	3938	90	10891	1653	215	226	
	3662	3662	3154	0	3587	1318	169	202	
	6196	6139	5045	57	24665.6	2226	430	430	
	3442	3440	2819	2	12752.9	1262	185	184	
	4258	4208	3607	50	4273.8	1559	153	178	
	3960	3958	3209	2	1625.6	1520	155	160	
	1866	1866	1632	0	4073.3	675	79	93	
	3701	3484	2981	217	5774.7	1283	174	196	
	1777	1777	1524	0	9352.6	543	95	96	
	3965	3663	2983	302	795.5	1307	183	184	
	2747	2737	2355	10	3505.2	941	111	133	

0 of 2341 selected

As shown in the figure, the population density is in the POPDENS_CY field, as noted in lesson 1. This attribute stores population per square mile for the current year, in this case census year 2015. Close the attribute table when finished.

2) Open the attribute table of block_groups.shp and scroll across its attributes.

The table has many of the same fields as the tracts.shp table, but there is no population density attribute. This won't be a problem because population density is just total population divided by area. Because you have total population, you need only the area of each block group, which is something that ArcGIS Pro can calculate automatically. So either shapefile satisfies your need: tracts already has population density, and you can derive it from block_groups.

You still need a definition of "densely populated." To keep it simple, you can call a neighborhood densely populated if its value exceeds that of the city of Los Angeles. The population density of Los Angeles as of the year 2015 was 8,474.7 people per square mile. Rounding down to an even number, use 8,500 people per square mile as your threshold.

3) In row 8 of the data requirements table, in the Defined As column, enter >= 8,500 per sq mi.

Threshold values

To find the population density of Los Angeles, as well as other threshold values for the analysis, we used online US Census Bureau data, especially the QuickFacts page at http://www.census.gov/quickfacts.

Define "lots of children"

Again, you'll look for attributes and then set a threshold.

1) Open the attribute table of the block_groups layer and locate the POP18UP_CY attribute.

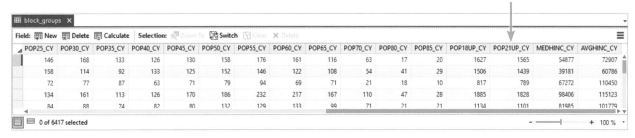

This is the population age 18 or older for the year 2015. If you define a child as a person under 18 (which is reasonable), you can subtract the values in this field (POP18UP_CY) from those in TOTPOP_CY to get the number of children.

Because neighborhoods vary in size and population, you can make more valid comparisons if you base your threshold on a ratio rather than an absolute number. In Los Angeles, 22.2 percent of the population is under 18. You'll therefore define a neighborhood as having "lots of children" if it meets or exceeds this value. (You can derive the percentage of children from your data using simple arithmetic.)

2) In row 9 of the data requirements table, in the Defined As column, enter >= 22%.

3) In the Attribute Data column, replace the entry "age" with age under 18.

Define "lower income"

Now look at your income attributes.

1) Open the attribute table of block_groups and scroll all the way to the right. The last two attributes might be income measures. To find out, you'll look at the metadata—not the item descriptions you looked at before, but the full data documentation.

2) Click the Project tab (on the far end of the ribbon) and click Options.

3) Click Metadata in the left column and change the Metadata style to North American Profile of ISO19115 2003.

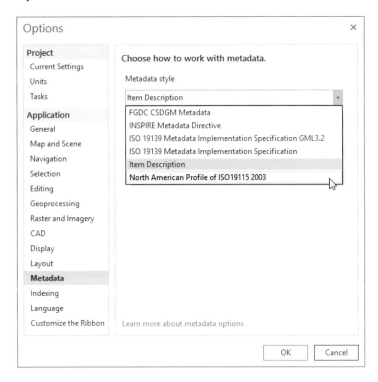

4) Click OK and return to the project using the button.

Changing the metadata style gives you access to the full set of metadata from the dataset. See the sidebar "Metadata" later in the lesson for more information.

5) If necessary, open the Project pane and browse to ParkSite > SourceData > Census.

6) Right-click block_groups.shp and click View Metadata.

The metadata is divided into three sections that can be expanded or collapsed. (They are expanded by default).

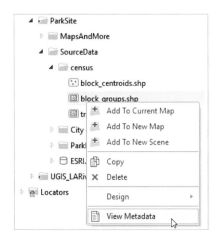

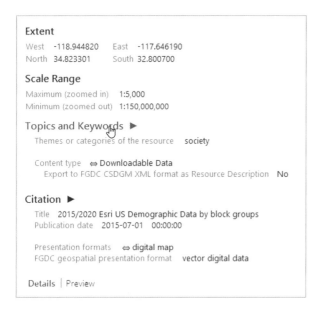

7) Click Topics and Keywords to collapse it. Right-click block_groups.shp and click View Metadata.

8) Collapse the next several headings (Citation, Citation Contacts, and so on) until you come to the Fields heading.

This heading contains the metadata that you're interested in.

9) Scroll through the Fields data.

10) Scroll down until you see the MEDHINC_CY and AVGHINC_CY fields and read the description.

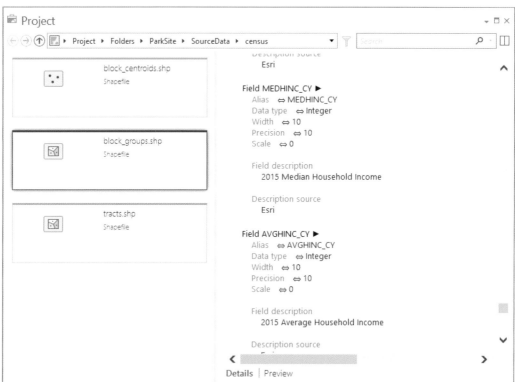

Note that MEDHINC_CY is described as 2015 Median Household Income and AVGHINC_CY is 2015 Average Household Income.

Metadata

Metadata is a description of what is known about a dataset. It serves two important purposes. First, it vouchsafes the integrity of data by explaining things such as how, when, and by whom the data was created. Second, it makes the data searchable. Metadata includes tags that identify essential properties of the data (for example, "rivers," "Los Angeles," and "2010") and other descriptions that make it possible to find specific datasets among large inventories of spatial data.

Metadata may be kept according to one of various official standards. Data created by government agencies, commercial data vendors, and many large enterprises typically conforms to one of these standards. Data created by small organizations or by individuals commonly does not. In ArcGIS, metadata can be displayed in a style that is suited to a particular standard. The default style is the Item Description, which displays a thumbnail image of the data and a small amount of important information. This style is suited to metadata that is not kept to an official standard. It can also be used to provide a filtered, summary view of metadata that is kept to an official standard. Anyone who creates and shares data should at least maintain metadata at the Item Description level.

To see the full metadata for a dataset that is kept to an official standard, you must change the metadata style in ArcGIS Pro, under Project Options. All the styles, apart from Item Description, are similar, and all afford access to the full set of metadata—no matter what standard they conform to—though they may present the information slightly differently.

Both are good possibilities. Median income is a statistical midpoint: it marks the value that half the households are above and half are below. You'll adopt this measure because it's less sensitive to extreme values. (A millionaire in a low-income neighborhood might significantly change the average income but not the median income.)

According to the US Census Bureau, the median household income for the city of Los Angeles for the years 2010–14 is $49,682. Rounding off, you'll call a neighborhood "lower income" if the median household income is $50,000 or less.

11) **In row 10 of the data requirements table, in the Defined As column, enter <= $50,000.**

12) **In the Attribute Data column, replace "income" with** median hh income.

Define "serving the most people"

Finally, you want to know which potential site serves the most people. Anyone can come to a park, so for this criterion you want to count all the people nearby, regardless of their demographic profile. The attribute you need for this element is total population, which you have in both the tract and block_ group feature classes.

You'll treat this guideline as a preference. If half a dozen sites meet your other requirements, you prefer those serving more people overall to those serving fewer. Eventually, this preference may have to be subjectively weighed against others. For example, which is better: a park closer to the river that serves fewer people or a park farther from the river that serves more people? (How much closer? How many more people?)

To define the size of the area served by a park, you'll apply the standard of easy walking distance discussed earlier, and say that a park serves anyone who lives within a quarter mile of it. "Serving the most people" therefore means having the largest population within a 0.25-mile radius.

1) **Open the block_centroids table and browse across its attributes.**

This table also has a total population attribute (POP2010). You can't use block centroids as your spatial data for neighborhoods—you need polygons rather than points—but you can conveniently use them to sum population. Given a distance of a quarter mile around the park, ArcGIS Pro can count the block centroids, or points, that fall within this distance and add their population values.

2) In row 11 of the data requirements table, in the Defined As column, enter <= 0.25 miles.

Your data requirements table should look like the figure.

#	REQUIREMENT	DEFINED AS	SPATIAL DATA	ATTRIBUTE DATA	DATASET	PREPARATION
1	land parcel		parcels			
2	vacant			land use		
3	a quarter acre or more			area		
4	within city limits		cities			
5	near LA River	<= 0.5 miles	rivers			
6	away from other parks	>= 0.25 miles	parks			
7	in a neighborhood		census unit			
8	densely populated	>= 8,500 per sq mi		population density		
9	lots of kids	>= 22%		age under 18		
10	lower income	<= $50,000		median hh income		
11	serving the most people	<= 0.25		population		
12	final map		political boundaries			
13	final map		roads			
14	final map		relief			
15	final map		imagery			

You can now state the problem in measurable terms that allow you to solve it with GIS tools. Someone might take issue with your interpretation of the city council's park guidelines, but that's fine—you'll always be happy to improve your methodology. For now, you can state your project analysis requirements as follows:

You want to locate a site for a new park on a land parcel that is

- vacant,
- a quarter acre or more in size,
- within the LA city limits,
- within 0.5 miles of the LA River (preferring closer sites),
- more than 0.25 miles from the nearest park,
- in a census unit in which
   - the population density is 8,500 or more people per square mile,
   - where at least 22 percent of the population is under 18 years old, and
   - where median household income is $50,000 or less, and
- considering that all other conditions are satisfied, the total population within a 0.25-mile radius is maximized.

You've formalized the data requirements in a table and confirmed that the essential data is available. In some cases, multiple datasets contain suitable features and attributes. In the next lesson, you'll compare these datasets and choose which ones to use in the analysis.

3) Close the Lesson2 map and any open tables.

4) Save your project.

5) Continue to the next lesson or close ArcGIS Pro. Save your changes if prompted.

Lesson

3 Choose the data

CHOOSING THE DATASETS TO USE

in the analysis is your first goal in this lesson. You came across a few situations in lesson 2 (with rivers, parks, and census units) in which different datasets stored similar features and attributes. How do you choose one dataset over another, which you'll be doing in exercise 3a?

Here are some considerations:

- One dataset may represent features with a more suitable geometry type for your needs (think of cities as polygons versus cities as points).
- Among datasets of the same geometry types, one may represent features with a more appropriate degree of detail.
- One dataset may be of more recent vintage than another. Or it may be more complete, including features that the other lacks.
- One dataset may be more accurate than another when viewed against imagery or other basemaps. Or it may have been created by a more authoritative source. Or it may conform better to another dataset you've already decided to use.
- One dataset may need less preliminary processing of spatial or attribute data to make it ready for analysis.

Your second goal is to choose a coordinate system for the analysis. In reality, that decision is often determined by prevailing standards in your organization, or by more or less default best choices for a given study area. For example, it makes sense for you to use the state plane coordinate system of 1983 because it is specifically designed to minimize spatial distortion for more than a hundred local areas (such as yours) within the United States. Although the choice of a coordinate system bears directly on your project, exercise 3b also contains background material about coordinate systems (what they are, why they matter, how they're managed) that is important to know for any GIS work.

Lesson Three road map

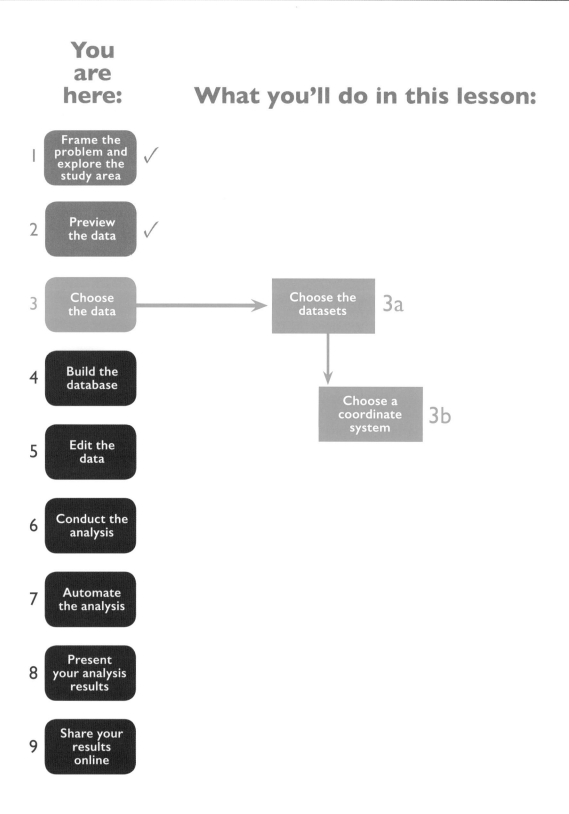

Exercise 3a: Choose the datasets

In this exercise, you'll add layers to ArcGIS Pro, compare them, and decide which datasets to use in the analysis. You'll add this information to the data requirements table you used in lesson 2.

Open the date requirements table

1) Using Windows Explorer, open the data requirements table you've been working on from UGIS\ParkSite\MapsAndMore.

Choose the parcel data

Parcels are the individual property boundaries from which your candidate park sites will eventually be selected.

In lesson 2, you previewed the Parcels shapefile and the Vacant Parcels table in the City of LA folder. Since you don't have any other parcel data, you will use these two datasets.

1) In row 1 of the table, in the Dataset column, enter Parcels.

2) In row 2, under Dataset, enter Vacant Parcels.

Recall that the Vacant Parcels dataset is a "stand-alone table." It does not include any polygons. You must join the Vacant Parcels table to the Parcels attribute table to see vacant parcels on the map. You'll do this table join—along with all your other data preparation—in lesson 4, but you'll make a note of it here.

3) In row 2 of the data requirements table, in the Preparation column, enter join table to Parcels.

You must also calculate parcel acreage to satisfy your third requirement.

4) In row 3, under Dataset, enter Parcels. Under Preparation, enter calculate area.

#	REQUIREMENT	DEFINED AS	SPATIAL DATA	ATTRIBUTE DATA	DATASET	PREPARATION
1	land parcel		parcels		Parcels	
2	vacant			land use	Vacant Parcels	join table to Parcels
3	a quarter acre or more			area	Parcels	calculate area
4	within city limits		cities			
5	near LA river	<= 0.5 miles	rivers			
6	away from other parks	>= 0.25 miles	parks			

Get started

1) Start ArcGIS Pro and open your LA River project. You'll work in ArcGIS Pro in this lesson and access data from the Project pane.

2) Make sure the Project pane is either open or visible as a tab on the right side of the ArcGIS Pro window. If necessary, open the Project pane on the View tab and click Project.

3) Insert a new map and name it Lesson3a in the Project pane.

Choose the city limits data

The requirement in row 4 is that the new park be within the city limits. You assumed this requirement meant that you'd need spatial data representing the boundary of Los Angeles (which you have), but actually, there's an even simpler solution.

1) In the Project pane, expand Folders > ParkSite > SourceData > City of LA folders and add VacantParcels.dbf to the Lesson3a map.

2) Open the table.

3) Find the CityCode field on the right side of the table. This field tells you which city each vacant parcel is recorded in.

4) Make sure you're scrolled to the top of the table. Right-click CityCode and click Sort Ascending. The first record is LA.

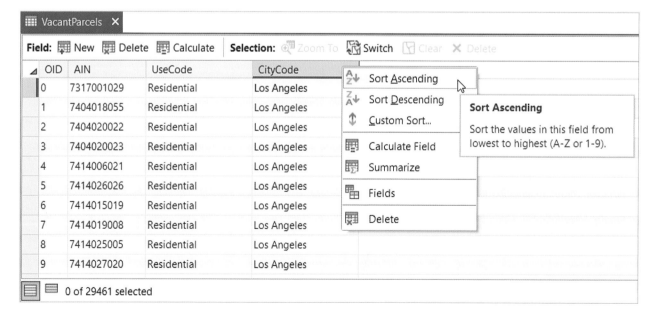

Lesson 3: Choose the data

5) Right-click CityCode again and click Sort Descending.

The value for CityCode doesn't change, which tells you that every record in the field has the same value. This means that all your vacant parcels—and consequently, all your potential park sites—are prequalified as belonging to Los Angeles. It turns out that you don't need a spatial boundary after all to guarantee that a parcel lies within the city limits.

If the Sort Ascending/Sort Descending command is dimmed, try double-clicking the CityCode field heading to sort the field. The arrow on the right of the field name will indicate if the field is sorted ascending ▲ or descending ▼.

6) Double-click the CityCode field heading again to toggle the sorting back to ascending.

7) Close the table.

8) In row 4 of the data requirements table, under Dataset, enter Vacant Parcels.

9) Also in row 4, under Spatial Data, delete cities. Under Attribute Data, enter city name.

Add and symbolize the river data

To analyze the distance of parcels to the river, you need spatial data representing the Los Angeles River. As you saw in lesson 2, you have two datasets that might work—the River feature class in the geodatabase or the LARiver shapefile. You'll compare these two datasets against an imagery basemap to see which dataset might be more appropriate.

1) In the Lesson3a map, change the basemap to Imagery.

2) In the Project pane, expand ESRI.gdb and then Hydro.

3) Drag River to the map display. The display should zoom to the extent of the River feature class, and you should see all the rivers in the LA area.

The only river you want to see on the map is the LA River. In lesson 1, you solved this problem with a definition query. Another approach is to symbolize the Los Angeles River and no other features.

4) **On the Appearance ribbon, click Symbology.**

One option to symbolize data is by category. This works best when the attribute of interest is a name, a description, or a number that isn't a quantity, such as a code or ranking. The Unique Values method assigns a different symbol to each unique value in the specified field.

5) **Click Unique Values.**

6) Make sure the Value Field drop-down list is set to NAME.

7) Click the Add Values button.

8) When prompted to generate the full list, click Yes.

9) Scroll down to Los Angeles River. (The list is alphabetical.) Click on it to highlight the row and click OK.

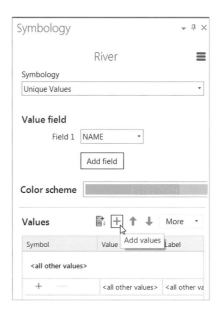

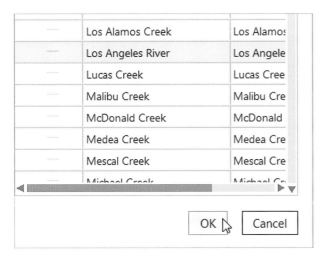

In both the Contents and Symbology panes, Los Angeles River is added and assigned a color from the current color ramp and a line width. By default, other features in the layer will get the symbol next to <all other values>. You don't want to see those features, however.

10) Click the More drop-down arrow, and click to clear the Show all other values check box. Selecting this option will turn off the other values in the layer.

11) Click OK.

12) Click the symbol patch next to Los Angeles River to open the Formal Line Symbol pane. Then go to the Properties tab.

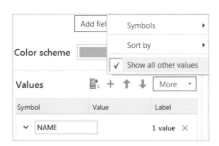

Exercise 3a: Choose the datasets   93

13) **Change the color to Apatite Blue. Change the line width to 2. Compare your settings to the figures and click Apply.**

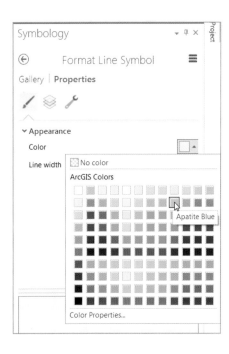
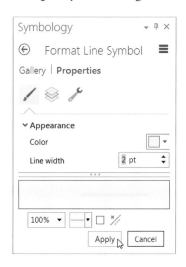

On the map, only the Los Angeles River is displayed using the symbology you defined.

14) **Close the Symbology pane.**

Symbolize the other river data

1) In the Project pane, under City of LA, drag LARiver.shp to add the layer to the map.

2) Under Contents, click the LARiver symbol patch to open Symbology.

3) Change the color to Cretan Blue and the line width to 2. Click Apply.

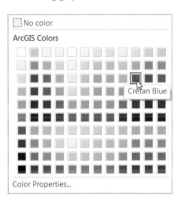

4) Close the Symbology pane.

Choose the river data

Even at this relatively small (zoomed out) scale, it's obvious that the two representations of the river have different spatial extents. One goes to the sea while the other stops abruptly halfway there. Now find out why.

1) In the Project pane, expand the MapsAndMore folder.

2) Drag LosAngeles.lyrx to the map.

3) **In the Contents pane, right-click the Los Angeles layer and click Zoom To Layer.**

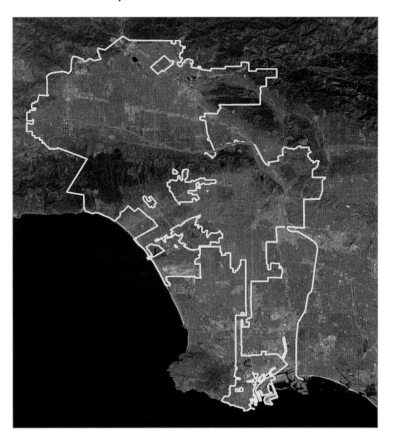

This is the layer file (the file of saved layer properties) that you created in lesson 1.

4) **Open the layer properties of the Los Angeles layer.**
5) **On the Layer Properties dialog box, click Definition Query.**

As reflected on the map, the symbol is a hollow yellow outline, and there is a definition query on Los Angeles.

6) **Click the Source tab to see the Data Source window.**

A layer file isn't a feature class—it doesn't store feature coordinates and attributes. Every layer in a map must point to a feature class on disk, and the same goes for a layer file. This layer file points to the City_ply feature class.

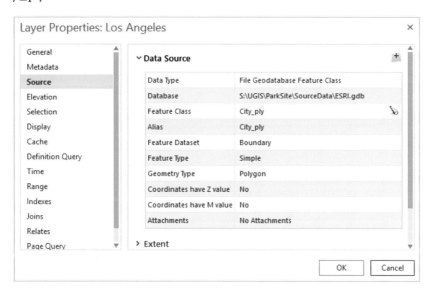

7) Close the Layer Properties dialog box by clicking OK.

You can see the likely explanation for the different extents of the two river layers: in the LARiver layer, the river stops at the city boundary. The City of Los Angeles, which provided the source data for this layer, doesn't need to maintain features beyond its own jurisdiction. (It's interesting that even a natural feature such as a river might be defined in part by administrative or political considerations.)

8) Zoom in on the source (the northwestern end) of the river. Set the map scale to 1:24,000.

9) Pan along the river and visually compare the two river representations. Go all the way to where the LARiver layer stops at the city boundary.

The two representations of the river aren't identical, but they're close. At large scales, such as 1:2,500, the river in LARiver is noticeably more sinuous and conforms more closely to the imagery, but for your analysis, either dataset is perfectly adequate. You'll use the LARiver dataset, which is already clipped to your area of interest.

10) In row 5 of the data requirements table, under Dataset, enter LARiver.

Exercise 3a: Choose the datasets   97

When you previewed this dataset in lesson 2, you saw that it was composed of 265 features. You can combine them into a single feature with a data processing operation called Dissolve. It will help get the data ready for lesson 6, in which you must create a half-mile buffer (proximity zone) around the river.

11) In row 5, under Preparation, enter dissolve.

#	REQUIREMENT	DEFINED AS	SPATIAL DATA	ATTRIBUTE DATA	DATASET	PREPARATION
1	land parcel		parcels		Parcels	
2	vacant			land use	Vacant Parcels	join table to Parcels
3	a quarter acre or more			area	Parcels	calculate area
4	within city limits			city name	Vacant Parcels	
5	near LA river	<= 0.5 miles	rivers		LARiver	dissolve
6	away from other parks	>= 0.25 miles	parks			

12) Save your project.

Add and symbolize the park data

To analyze the location of candidate sites in respect to existing parks, you need spatial data representing parks. You'll compare the Parkland dataset you used in lesson 1 to the Parks dataset you previewed in lesson 2.

1) In the Contents pane, turn off the River layer and click its side arrow to collapse it.

▶ Leave the LARiver layer turned on.

2) In the Project pane, under ESRI.gdb, expand Landmark.

3) Drag Parkland to the map.

4) In the City of LA folder, drag Parks.shp to the map.

The order of layers in Contents should look like the figure. Your symbology for the parks layers may be different.

98   Lesson 3: Choose the data

5) Under Contents, right-click the symbol patch for the Parks layer to open the color palette. Change the fill color to Chrysoprase.

6) Right-click the Parkland symbol patch and change its fill color to Tzavorite Green.

7) Select the Parks layer and click the Appearance tab. In the Effects group, change Layer Transparency to 50%.

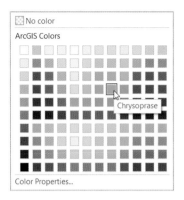

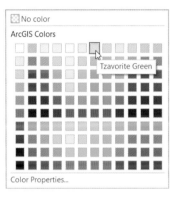

The symbology makes it clear where the two layers agree and disagree on the representation of features.

Exercise 3a: Choose the datasets

Examine a park in two layers

You can start with a close look at an example park. Your map should still be zoomed in to where the river crosses the city limits. The scale should still be 1:24,000.

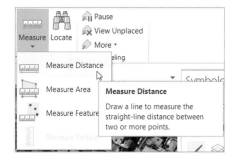

1) On the Map tab, in the Inquiry group, click the Measure tool and click Measure Distance.

2) Click the Choose Options button ▾, click Distance Units and Miles, and if necessary, click to clear Feet.

You're going to follow the river north about 2.25 miles.

3) Click at the end of the river to start the measurement. Move your pointer north along the river (you don't have to drag).

4) As you move the pointer, the Segment length value changes in the Measure dialog box.

5) When you reach the top of the map display—assuming your measurement is still less than 2.25 miles—press and hold the C key on your keyboard.

The pointer changes to the hand icon 🖐.

6) Pan north and release the C key to continue the measurement.

7) When the length of the measurement reaches about 2.25 miles, double-click to end the measurement.

Half a mile due east of the end of your measurement is a park.

8) Switch back to the Explore tool on the Map tab and click on the center of the park where the two park layers overlap to identify the park.

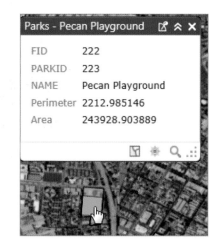

Its name is Pecan Playground. The attributes in the pop-up window come from the Parks layer, not from Parkland, because of the layer order in the Contents pane.

By default, the topmost layer is identified. You can choose other layers from the drop-down list under Explore.

9) **In the lower-right corner of the Identify window, click the Zoom tool, which will zoom to the feature Pecan Playground.**

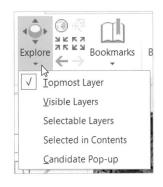

The map zooms in on the park.

10) **Create a new bookmark named Pecan Playground.**

11) **Close the Identify window and the Measure dialog box.**

12) **Turn the Parkland layer off and on a few times.**

Neither feature conforms to the image: both encroach, for example, on the surrounding streets. The Parks feature seems better, however, because it includes the swimming pool and the play area at the north end of the block. You'll come back to this place in lesson 5 to do some spatial editing.

13) **Under Bookmarks, click Dodger Stadium to zoom to it. Your view is now centered on Dodger Stadium. Compare the Parkland and Parks layers.**

Choose the park data

You'll follow the river upstream and look at some more parks to help you reach a decision.

1) **In Contents, drag the Los Angeles layer directly underneath the LARiver layer.**

2) **Set your map scale to 1:36,000. Start panning northwest along the river's course.**

You'll soon come to Griffith Park, the vast park at the river's elbow. Both the river and the city limits run along the park's edge. Notice that to the east (Glendale) and north (Burbank) of the city limits, there are features from the Parkland layer, but no corresponding Parks features. This is another case of jurisdiction. The City of Los Angeles, which supplied the Parks dataset, doesn't maintain data for parks in other cities.

This difference in managing data raises a question. The new park must be a quarter mile away from existing parks: Does that mean any park in any city or just any park in Los Angeles? You'll interpret it to mean any park in any city, because, by and large, there are no residency requirements for park use. On that basis, you should choose the Parkland feature class for your analysis. You don't know if it's more or less accurate on the whole (you only examined one park), but it covers a part of your area of interest that the Parks feature class doesn't.

3) In row 6 of the data requirements table, under Dataset, enter Parkland.

Add more park data

In lesson 2, you previewed the NewParks feature class and Vista Hermosa Park, which has been completed but is not in NewParks.shp. You should find out if these features are included in the Parkland layer.

1) In the Project pane, expand the ParkData folder and add NewParks.shp to the map.

2) In Contents, drag the NewParks layer directly above the Parks layer.

3) Right-click the symbol patch for the NewParks layer and click Macaw Green.

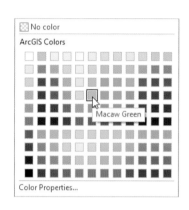

4) Zoom to the NewParks layer.

▶ How? In the Contents pane, right-click the layer and click Zoom To Layer.

5) **Turn the NewParks layer off and on a couple of times.**

You can see parks on the basemap imagery, but they're not represented in the Parkland layer. Another data preparation task in lesson 4 will be to add these missing parks into the Parkland feature class.

6) **Zoom to Vista Hermosa Park using the bookmark you created in lesson 2.**

There's no Parkland feature here, either, or you'd be looking at it. In lesson 5, you'll create a park feature in this location.

7) **In row 6 of the data requirements table, under Preparation, enter edit features.**

Choose the census data

In lesson 2, you decided that you would represent neighborhoods with census units. Your two choices are block groups or tracts. Block groups meet all your attribute needs, either directly or indirectly. Because they are smaller than tracts, block groups also portray demographic patterns in more detail.

1) **In rows 7 through 10 of the data requirements table, under Dataset, enter block groups.**

Population density and age under 18 attributes don't exist as such in the block_groups table, but you can derive them from other attributes.

2) **In rows 8 and 9, under Preparation, enter calculate.**

The last analytical requirement is to count how many people live within a quarter mile of each proposed park site. In lesson 6, you'll do this by creating quarter-mile rings (buffers) around the proposed sites, counting the census block points in each ring, and summing their populations.

3) **In row 11, under Dataset, enter block centroids.**

#	REQUIREMENT	DEFINED AS	SPATIAL DATA	ATTRIBUTE DATA	DATASET	PREPARATION
1	land parcel		parcels		Parcels	
2	vacant			land use	Vacant Parcels	join table to Parcels
3	a quarter acre or more			area	Parcels	calculate area
4	within city limits		cities	city name	Vacant Parcels	
5	near LA river	<= 0.5 miles	rivers		LARiver	dissolve
6	away from other parks	>= 0.25 miles	parks		Parkland	edit features
7	in a neighborhood		census unit		block groups	
8	densely populated	>= 8,500 per sq mi		population density	block groups	calculate
9	lots of kids	>= 22%		age under 18	block groups	calculate
10	lower income	<= $50,000		median hh income	block groups	
11	serving the most people	<= 0.25 miles		population	block centroids	

Choose the final map data

The last rows of the table list data that you may need for cartographic reasons in the final map.

Row 12 lists political boundaries. Your source data has feature classes of cities, counties, and states. Until you design your final map, you won't know for sure, but you can foresee a likely need for city and county boundaries, but not for state boundaries.

1) **In the Project pane, under ESRI.gdb, expand Boundary.**

In lesson 2, you previewed City_ply and saw that its extent covered the entire United States. That's more data than you'll need. You never previewed the County feature class, so do that now.

2) Under Boundary, drag the County feature class to your map.

3) Zoom to the layer and look at the attribute table.

Note that the attributes include both the county and state names along with census attributes similar to those you've seen in other layers.

4) Turn off the County layer and close its attribute table.

5) In row 12 of the data requirements table, under Dataset, enter City_ply. Press Enter to start a new line and enter County.

6) In row 12, under Preparation, enter reduce extent next to both City_ply and County.

In row 13, in the Spatial Data column, you have a need for roads listed.

7) In the Project pane, under ESRI.gdb, expand Transport and drag Mjr_rd to your map.

The data covers the Greater Los Angeles Area, which is appropriate for your map, but you also know that roads are available on the Topographic basemap.

8) Change the basemap to Topographic using the Basemap button on the Map tab, in the Layer group.

9) Zoom in so you can see the difference between the Mjr_rd layer and the basemap. Optionally, switch off between the Imagery basemap and the Topographic basemap to compare accuracy.

The Topographic basemap is more detailed and more accurate. The layer is also updated automatically for you by Esri, so your maps will stay current into the future. The roads on the basemap are also already cartographically styled nicely with labels so it will be a lot easier to work with. One consideration with basemaps is that they are essentially an image—you can't query or select from them, and you can't use them as the input for tools. Your project, however, doesn't include roads in the criteria; you simply want roads for reference and cartographic purposes. For these reasons, you will use the Topographic basemap to display roads rather than using the Mjr_rd dataset.

10) In row 13 of the data requirements table, under Dataset, enter basemap.

Relief and imagery are also available as online basemap layers.

11) In rows 13, 14, and 15 of the data requirements table, under Dataset, enter basemap.

#	REQUIREMENT	DEFINED AS	SPATIAL DATA	ATTRIBUTE DATA	DATASET	PREPARATION
1	land parcel		parcels		Parcels	
2	vacant			land use	Vacant Parcels	join table to Parcels
3	a quarter acre or more			area	Parcels	calculate area
4	within city limits		cities	city name	Vacant Parcels	
5	near LA river	<= 0.5 miles	rivers		LARiver	dissolve
6	away from other parks	>= 0.25 miles	parks		Parkland	edit features
7	in a neighborhood		census unit		block groups	
8	densely populated	>= 8,500 per sq mi		population density	block groups	calculate
9	lots of kids	>= 22%		age under 18	block groups	calculate
10	lower income	<= $50,000		median hh income	block groups	
11	serving the most people	<= 0.25 miles		population	block centroids	
12	final map		political boundaries		City_ply / County	reduce extent / reduce extent
13	final map		roads		basemap	
14	final map		relief		basemap	
15	final map		imagery		basemap	

Save your work

All the datasets you need for the analysis have been specified, so you can save and close your work.

1) Save and close the data requirements table.
2) Zoom to the Los Angeles layer.
3) Close the Lesson3a map.
4) Save your project.
5) If you are continuing to the next exercise now, leave ArcGIS Pro open; otherwise, close ArcGIS Pro.

In the next exercise, you'll take a step back from the project. Analysis operations take place within a particular coordinate system, and choosing that system is an important part of setting up the project. To make a good choice, you should have some background knowledge of what coordinate systems are and how they're managed in ArcGIS Pro.

Exercise 3b: Choose a coordinate system

A coordinate system is a mesh of perpendicular intersecting lines superimposed on a surface. The point of intersection of any two lines is a unique location, which can be specified with two values (a coordinate pair). The values are measurements from a given reference point in a given unit of measure. The unit of measure may be an angular unit, such as degrees, or a length unit, such as feet or meters.

Coordinate systems can be applied to any surface, but you're interested in the surface of the earth. If you represent the earth with a spherical model, such as a globe, you have a curved surface to deal with. If you represent it with a flat model, such as a map, the surface is flat.

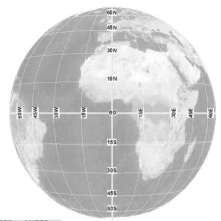

A geographic coordinate system is applied to round or roundish earth models. Geographic coordinate systems use an angular unit of measure because angles are better for measuring locations on a curved surface. A projected coordinate system is applied to flat earth models. Projected coordinate systems use a length unit of measure.

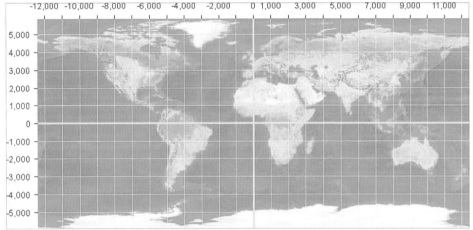

Every usable spatial dataset has a coordinate system. The locations of its features are specified by coordinates that are correct within its own framework but that would be wrong or absurd in another system. Fifty degrees of arc isn't the same as 50 meters, which isn't the same as 50 feet. Knowing the coordinates of a point doesn't tell you where the point is on the earth unless you also know what the coordinate system is.

There are many different geographic and projected coordinate systems. As long as two datasets use the same system, their features will align correctly on a map. Datasets with different systems must be reconciled. This basically means that one dataset's coordinate system must be mathematically converted, or projected, to the other. This can be done with data processing tools. ArcGIS Pro does it automatically when you add layers to a map. What ArcGIS Pro reconciles, however, are the coordinate systems of the layers in the map, not the coordinate systems of the datasets that the layers point to. This reconciliation of coordinate systems in a map is called "on-the-fly" projection.

In this exercise, you'll see how ArcGIS Pro manages on-the-fly projection of coordinate systems. You'll also look at how map display and map measurements change with different geographic and projected coordinate systems. Finally, you'll select a coordinate system for your analysis project.

Add data and check the coordinate system of a dataset

Every spatial dataset has a coordinate system. Now you'll see how to find out what it is.

1) If necessary, start ArcGIS Pro and open your LARiver_ParkSite project. Create a new map and rename it Lesson3b.

2) In the Project pane, under ESRI.gdb, expand Boundary and drag the State feature class to the Lesson3b map.

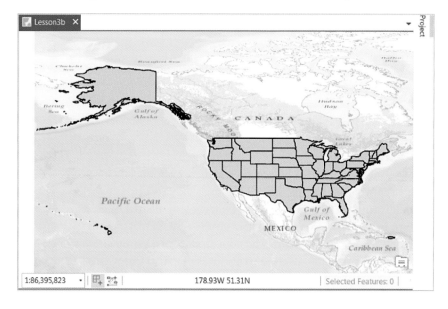

3) In the Layer Properties dialog box of the State layer, click Source. Scroll down to Spatial Reference and expand the group.

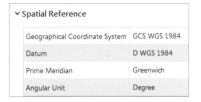

The window displays the current coordinate system: GCS WGS 1984.

"GCS" stands for geographic coordinate system. "WGS_1984" is the World Geodetic System of 1984, also called WGS84.

Under the name of the coordinate system are details giving the essential properties of the system:

- Its angular unit of measure (degrees of angle)
- Its prime meridian (Greenwich)
- Its datum (D_WGS_1984)

WGS_1984 is a geographic coordinate system. Geographic coordinate systems, and datasets that use them, are said to be "unprojected." If you checked the other feature classes in ESRI.gdb, you'd find that they all have this same geographic coordinate system.

4) **Close the Layer Properties dialog box.**

5) **Hover over the map.**

At the bottom of the application window, the coordinates of the pointer are reported in decimal degrees. The first coordinate (longitude) tells you the angular position east or west of the prime meridian. The second coordinate (latitude) tells you the angular position north or south of the equator.

6) **In the Contents pane, double-click Lesson3b to bring up Map Properties (or right-click Lesson3b and click Properties).**

7) **In the Map Properties dialog box, click Coordinate Systems. Confirm that the coordinate system is WGS 1984, the same as the State layer.**

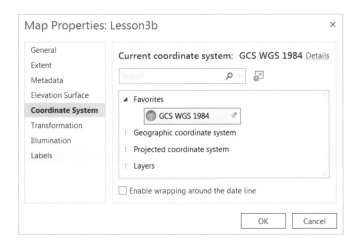

A map is a container for layers in a project. You can insert more maps to manage separate sets of data within your project. Why would you want to do that? It's sometimes a matter of layout: you may want to print a single map sheet that contains a main map, an overview map, an inset map, or some other combination of views. (You'll see how maps work in layouts in lesson 8.)

Exercise 3b: Choose a coordinate system 109

One crucial function of a map is to enforce the spatial alignment of the layers it contains. When you first add a layer to the map, it adopts that layer's coordinate system as a standard. Any layers added thereafter are converted (projected) by ArcGIS Pro into that same system. This process of on-the-fly projection happens automatically as new layers are added.

8) **Close the Map Properties dialog box.**

Project data on the fly

To see how on-the-fly projection works, you'll insert a second map, add a layer that has a different coordinate system from the State layer, and then copy the State layer to the new data frame. To see the content of both data frames at the same time, you must open another map.

1) **On the Insert tab, add a second map and name it Lesson3b Web Mercator.**

2) **Reposition Lesson3b Web Mercator on the right side of the Lesson3b map by dragging the tab of Lesson3b Web Mercator to the center of the map.**

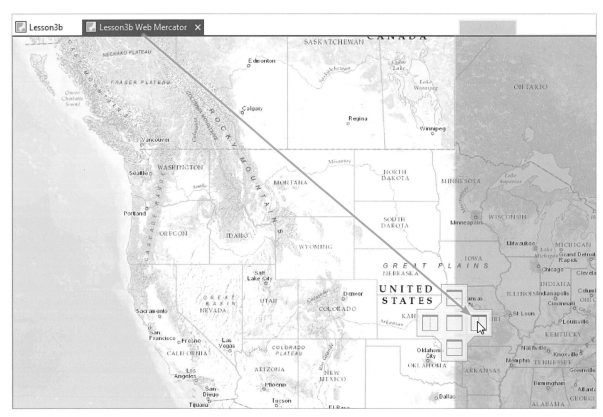

110   Lesson 3: Choose the data

When you release the mouse button, you'll have the two maps side by side. Adding a second map is usually done to add a new set of data layers in a new area of interest. In this case, the second map will allow you to compare two different map projections side by side.

3) On the View tab, click Link Views, and then on the drop-down menu, click Center And Scale.

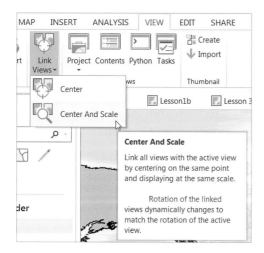

4) Open Map Properties for the Lesson3b Web Mercator map.

5) Click Coordinate System. Note that it is WGS 1984 Web Mercator (auxiliary sphere). This is the projection of the Topographic basemap. Close the Map Properties window.

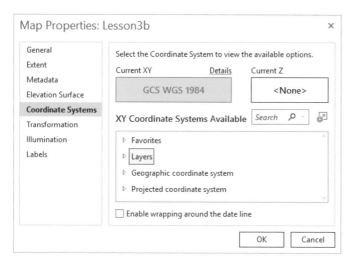

Exercise 3b: Choose a coordinate system 111

6) In the Contents pane of the Lesson3b map, right-click the State layer and click Copy.

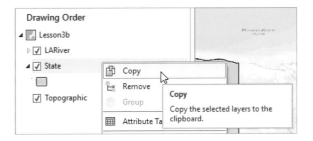

7) Click the Lesson3b Web Mercator tab to make it active, and then right-click Lesson3b Web Mercator in the Contents pane to paste the State layer under Contents.

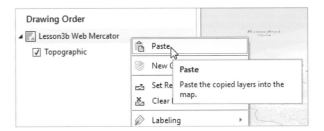

8) Examine the coordinate system of the Lesson3b Web Mercator map. Notice it changed to the projection of State: WGS 84. The two maps are identical but displayed in different projections.

With this new projection in the Lesson3b Web Mercator map, the basemap, which is in WGS 1984 Web Mercator (auxiliary sphere), is now being projected on the fly to WGS 84.

9) Change the coordinate system of the Lesson3b Web Mercator map back to WGS 1984 Web Mercator (auxiliary sphere). Click Projected coordinate system > World > WGS 1984 Web Mercator (auxiliary sphere). Click OK.

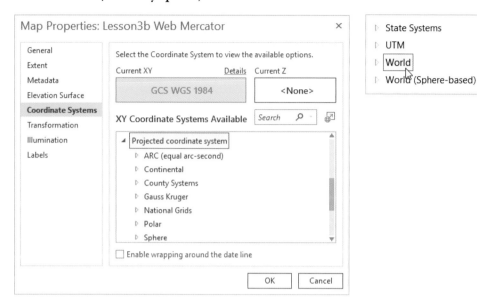

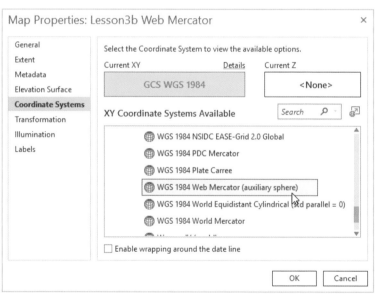

10) Compare the map appearance, and notice especially the shape of Alaska.

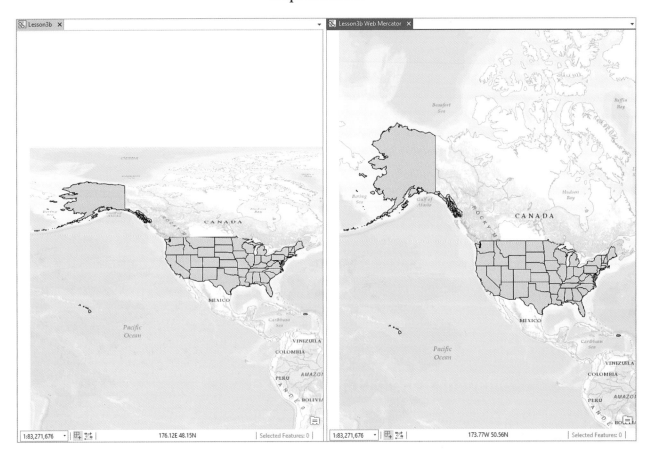

Make an area measurement

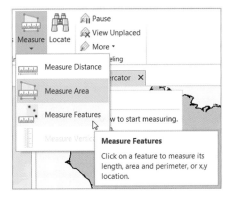

Coordinate systems change measurements as well as appearances. The Mercator projection you're using now is a conformal projection, which means that it shows shapes correctly. (Look at Alaska on a globe, and you'll see this is true.) On the other hand, it distorts area measurements—quite significantly in extreme latitudes.

1) Zoom in on Alaska—it doesn't matter how far exactly. Because the two maps are linked, both will zoom together.

2) On the Map tab, click the Measure tool and click Measure Features.

3) Under Options for the Lesson3b Web Mercator map, change the area Units and Mode. Click Area Units > Square Miles. Click Square Feet to turn it off. Then click Mode > Planar.

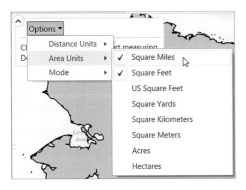
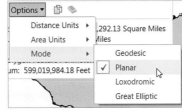

4) Under Options for the Lesson3b map, change the area Units and Mode. Click Area Units > Square Miles. Click Square Feet to turn it off. Then click Mode > Geodesic.

5) Click anywhere inside one of the two maps. Leave the Measure dialog box open.

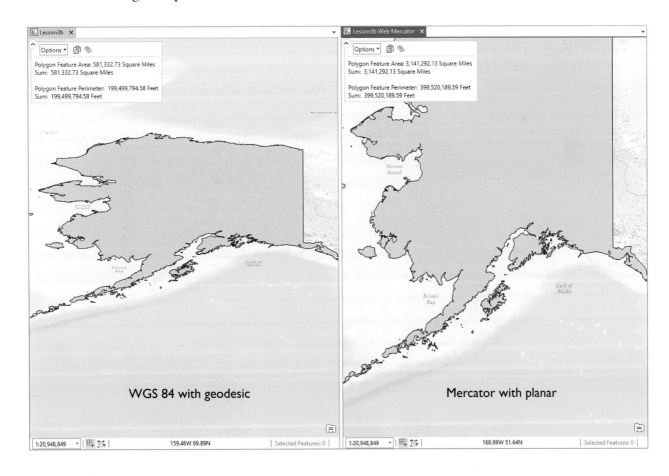

Exercise 3b: Choose a coordinate system   115

6) Open the State layer attribute table.

7) Select the record for Alaska. (It should be the second one.) Scroll all the way to the right to see the SQMI attribute.

The value in the table, which is correct, is 581,369 square miles. (You can find other values cited in reference sources, but they're similar to this one.) That means that the map measurement is an exaggeration by a factor of more than five.

8) At the top of the table window, click the Clear button. Close the table.

The Measure tool is doing its job correctly: it has measured the area of Alaska as distorted by the Mercator projection. (The Mercator projection has its virtues, but making area measurements in high latitudes is not one of them.)

Change the map's coordinate system

You can change the map's coordinate system whenever you want—before or after adding layers.

1) Open the map properties of the Lesson3b Web Mercator map.

2) Click Coordinate System. Change the projection from WGS_1984_Web_Mercator_Auxiliary_Sphere to Alaska Albers Equal Area Conic, which is found under the Projected coordinate system > Continental folder > North America folder.

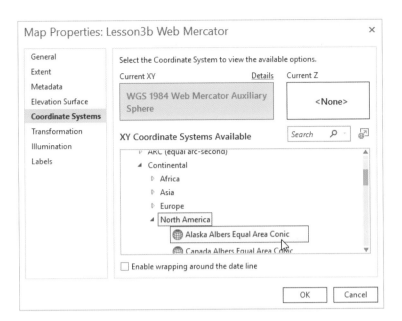

This system, unlike the Mercator, represents area (the size of features) correctly.

The new coordinate system is applied to the map. Both the basemap layer and the State layer are projected (reprojected, if you like) on the fly.

If you measured the area of Alaska again. your map should look like the figure.

Choosing a coordinate system

A suitable coordinate system for your project depends largely on your area of interest, and may also depend on whether you want to preserve a particular spatial property, such as shape or area, without distortion. A few systems are common standards. For example, web-based maps typically use the Web Mercator coordinate system. Local-area maps within the United States are frequently based on the state plane system, while local-area maps around the world (including the United States) often use the Universal Transverse Mercator (UTM) system.

On the Map Properties Coordinate System tab, you can search for coordinate systems covering only your area of interest. (This doesn't mean all the results are appropriate for your map. Systems designed for world maps are displayed with every spatial filter but may not be a good choice for a local map.) You can also enter text [california x ▾] to find coordinate systems that include the text as part of their name. This is often a good way to find systems specifically suited to your area of interest.

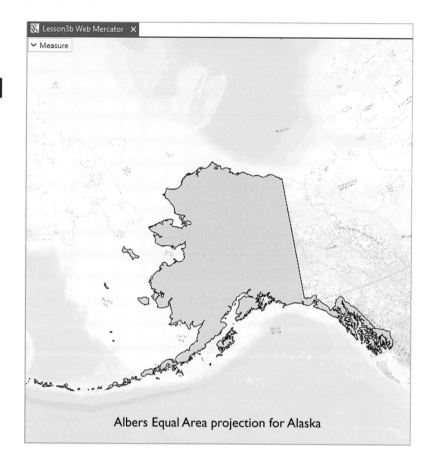

Albers Equal Area projection for Alaska

Coordinate systems are organized in folders. The two main folders are Geographic Coordinate Systems and Projected Coordinate Systems. Both folders contain several subfolders. Both the Web Mercator Auxiliary Sphere and Alaska Albers Equal Area Conic coordinate systems are in the Projected Coordinate Systems folder.

The more you learn about coordinate systems, the more useful this folder structure becomes. At the outset, it can be a bit overwhelming.

For more information, see the sidebars on "Geographic coordinate systems" and "Projected coordinate systems."

Geographic coordinate systems

A geographic coordinate system (GCS) is based on a spheroidal model of the earth. Sometimes a perfect sphere is used, but usually a slightly squashed "oblate spheroid" is used to reflect the fact that the earth bulges at the equator and is flattened at the poles.

Sphere   Oblate spheroid (much exaggerated)

The reference lines of a geographic coordinate system are parallels and meridians. Parallels are lines that circle the globe parallel to the equator. Meridians are lines perpendicular to the equator that converge at the poles. By convention, the origin of the system (its 0,0 coordinate) is the intersection of the equator and the prime meridian, the meridian passing through Greenwich, England.

Geographic coordinates, commonly called *latitude-longitude* values, are measurements of angle, not distance. Angles are a constant unit of measure on a sphere, whereas distances are not (because meridians converge).

Angle measurements are usually expressed in degrees, minutes, and seconds. A degree has 60 minutes; a minute has 60 seconds.

Latitude is angular position north or south of the equator. The equator is 0° latitude, the North Pole is 90° north, and the South Pole is 90° south.

Longitude is angular position east or west of the prime meridian. The prime meridian is 0° longitude. Its anti-meridian (on the other side of the world) is both 180° east and 180° west.

A latitude-longitude pair defines a unique position on the earth's surface.

The unique location of Dodger Stadium, for example, would be written like this: **34°4′26″N, 118°14′27″W** and spoken like this:

"34 degrees, 4 minutes, 26 seconds north latitude; 118 degrees, 14 minutes, 27 seconds west longitude."

For computer calculations, these values are converted to decimals. The location of Dodger Stadium in "decimal degrees" is 34.073, –118.24. The minus sign is used for west longitude and south latitude.

The fact that there are many different geographic coordinate systems is a source of trouble for GIS users. What makes two systems different is disagreement about the exact latitude-longitude values of particular locations. Why is there disagreement about that? In simple terms, it's because different spheroid models of the earth have been developed over time by different earth scientists using different technologies. Changing the shape or size of the model ends up changing the coordinates of points on the surface—usually not very much, but sometimes, in sensitive applications, enough to be of concern. This issue is taken up in more detail in the "*Datums*" sidebar later in this lesson.

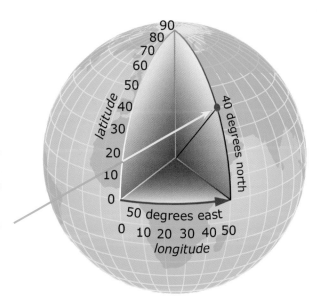

Exercise 3b: Choose a coordinate system

Projected coordinate systems

The earth's surface can be modeled very well on a spheroid, but not as well (except over small areas) on a plane. To make a map, you more or less have to flatten a sphere, which is like squaring a circle, only harder. It can't be done without radically adjusting the spatial properties and relationships of features on the surface: their shapes, sizes, and relative distances and directions.

The name for any such radical adjustment is a *map projection*. A projection is a mathematical formula (there are lots of different ones) for translating the world into flat space. All map projections introduce spatial distortion. They are variously designed to minimize certain kinds of distortion or to distribute it in certain ways over the map surface. Some projections correctly preserve feature shapes but distort their areas. Some preserve areas but distort shapes. Some compromise. Some have special properties, such as keeping true distance measurements from a single point to all others, or ensuring that courses of constant compass bearing are plotted as straight lines. The smaller the area being mapped, the less distortion there is of any spatial properties. Areas up to medium-size countries (say about the size of Nigeria or Bolivia) can be mapped with distortion so low as to be insignificant for most purposes.

A projected coordinate system consists of a map projection, a length-based unit of measure, an origin point for measurements, and other parameters, such as standard lines that define the distortion pattern on the map. Manipulating these parameters is what allows you to customize a coordinate system for a specific area of interest. Because a projection is applied to a particular spheroid and its definition of latitude-longitude values, a projected coordinate system also includes, or presupposes, a geographic coordinate system.

The idea of projection includes both going from a geographic (unprojected) system to a projected system, and going from one projected system to another (sometimes called *reprojection*). To go from one projected system to another, ArcGIS Pro undoes the map projection, goes back to the underlying geographic coordinate system, and applies a new projection to it. ArcGIS Pro stores thousands of map projection formulas and can run these calculations quickly.

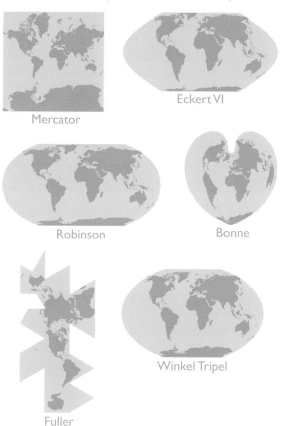

Some common world map projections (and a couple of uncommon ones).

Mercator

Eckert VI

Robinson

Bonne

Fuller

Winkel Tripel

When you add an unprojected dataset to a map (that is, a dataset that stores feature coordinates as latitude-longitude values), the data still must be projected in some sense to be viewed as a flat map on your monitor. In ArcGIS Pro, this default "pseudoprojection" has the display properties of a map projection (specifically, the Plate Carrée), but none of the other properties or parameters of a projected coordinate system.

Geographic pseudoprojection

Measure Alaska again

Equal-area projections (of which the Albers Equal Area Conic is one) preserve the spatial property of area, or size. The trade-off is that they don't represent shapes correctly. Because this projection is customized for Alaska, however, most distortion of all types is pushed outside the area of interest.

1) On the Map tab, click the Measure tool, and click Measure Features to measure Alaska again using this new projection.

The area is given accurately as 581,333 square miles.

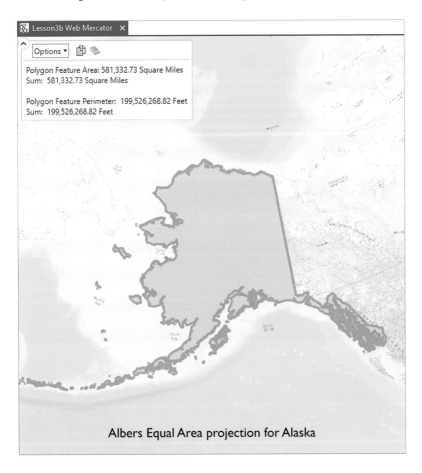

Albers Equal Area projection for Alaska

2) Switch from the Measure tool back to Explore.
3) Zoom to the full extent.

This coordinate system is obviously unsuitable for a world map. It doesn't show the whole world, for one thing. The price for mapping Alaska with very low distortion is that places such as Australia and South America are severely distorted—which is fine, as long as you keep your map zoomed in to Alaska.

Now you can summarize how things stand with the layers in this data frame:

- The geographic coordinate system of the State layer is GCS_WGS_1984. It has no projected coordinate system.
- The projected system of the basemap layer is Web_Mercator. Its underlying geographic system is GCS_WGS_1984.
- The map has been set to the Alaska Albers Equal Area Conic projected system. Its underlying geographic system is GCS_North_American_1983.
- The State and basemap layers have been projected on the fly into the Alaska Albers system.

No matter how often you change coordinate systems, ArcGIS Pro keeps the data in alignment.

4) **Change the projection back in Lesson3b Web Mercator to WGS 1984 Web Mercator (auxiliary sphere) (under Projected > World) from Alaska Albers Equal Area Conic.**

5) On the View tab, click Link Views to turn off link views.

6) Close the Lesson3b Web Mercator map.

▶ Reminder: it is still in the project under the Maps folder.

Add another layer

Now you can get back to your true area of interest.

1) **Copy the LARiver layer from Lesson3a and paste the layer into Lesson3b.**

2) **Zoom to the LARiver layer and turn off the State layer.**

A coordinate system meant for Alaska (or the world) isn't appropriate for Southern California. Local mapping needs across the United States are served by a system called the state plane coordinate system. This isn't a single coordinate system for the entire country, but a patchwork of systems, each of which covers a state or part of a state. Together, they assure that for whichever part of the country you want to map, you get uniformly low distortion. California is divided into six state plane zones, with Los Angeles falling in zone 5.

California Zone 5

Exercise 3b: Choose a coordinate system   123

The state plane coordinate system divides states into zones and applies a locally optimized coordinate system to each zone. Vertical zones are based on the Transverse Mercator projection. Horizontal zones are based on the Lambert Conformal Conic projection.

Change the map's coordinate system

You'll change the map's coordinate system again, this time to State Plane California Zone 5.

1) **Open the map properties of Lesson3b.**

2) **On the Coordinate Systems tab, expand the Layers folder.**

Within this folder are the three native coordinate systems of the layers in the map.

3) **Expand the coordinate system NAD 1983 StatePlane California V FIPS 0405 Feet.**

This is the coordinate system used by the LARiver layer. (Strictly speaking, it is the coordinate system of the LARiver shapefile.)

4) **Click the layer name to highlight it.**

This resets the current coordinate system of the data frame. The state plane coordinate system for California zone 5 is a projected coordinate system based on a Lambert Conformal Conic map projection.

Its underlying geographic coordinate system is North American 1983. This is different from the WGS 1984 system used by both the State and basemap layers.

5) **Click OK.**

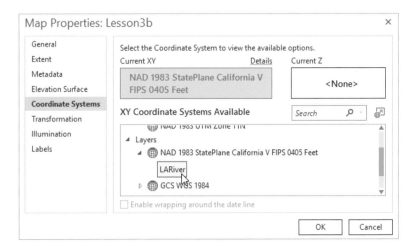

Your map should look like the figure. The state plane coordinate system for California zone 5 is the one you'll use for your analysis.

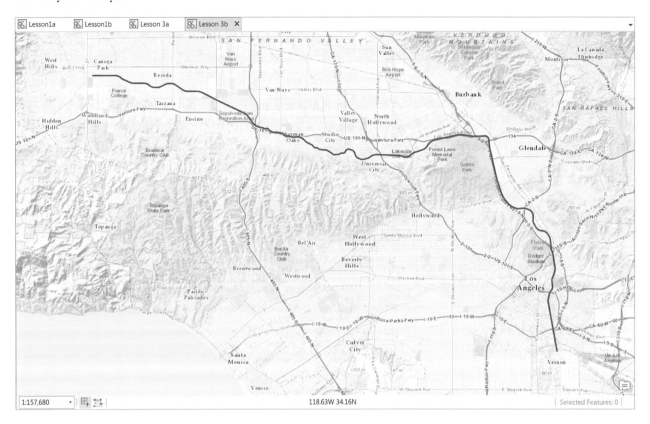

6) Close all the Lesson3 maps and any open tables.

7) Save your project.

8) Continue to the next lesson or close ArcGIS Pro. Save your changes if prompted.

In lesson 4, you'll resume work on the project by taking care of the preparation tasks you listed in the data requirements table.

For more information on datums, see the sidebar.

Datums

A geographic coordinate system is defined by three things: an angular unit of measure (usually degrees), a prime meridian (usually Greenwich), and a datum. The datum is the part that gives people trouble. To understand it, start with the shape of the earth.

The earth isn't a perfect sphere, or even a mathematically regular spheroid. It's a lump with an uneven shape owing to different concentrations of mass (and therefore unequal gravity) over its surface. In addition, it has topographic features such as mountains and valleys.

When the spatial positions of features are determined—as was formerly done by survey, and is now mostly done by satellite—they are first determined on the earth's surface. These raw measurement values are then mathematically "leveled" to a geoid. A geoid is the (still gravitationally lumpy) shape that the earth would have if it was covered by a mean sea level surface—in other words, if it had no topography.

The shape of the geoid, however, is too complex to be a working model. So the next step is to move the measurements from the geoid to a spheroid: a model with a regular, nonlumpy shape.

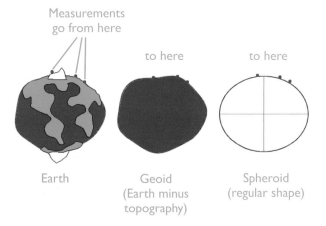

That's where the datum comes in. The datum is two things: first, it's a chosen spheroid, which could be WGS 1984, GRS 1980, Clarke 1866, Bessel 1841, or a number of others. (The world is standardizing on the GRS 1980 spheroid but isn't all the way there yet.) Second, it's a mathematical orientation, or "fit," of the geoid to the spheroid. In the transfer of measurements from geoid to spheroid, some error will be introduced because the lumps must be smoothed out.

How that error is distributed is the "fit." One approach is to make the fit really good for one part of the world, such as North America, and not to worry about the rest. That's a local datum. It's designed to maintain high accuracy for measurements over a limited area. The other approach is to average the error over the whole surface. That's an earth-centered datum. It's designed to maintain high accuracy for the world as a whole.

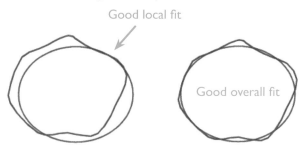

When two geographic coordinate systems are different, it's usually because the datums are different (which, in turn, is either because the spheroids are different or the fit is different). When you get a coordinate system warning in ArcGIS Pro, one possibility is to ignore it and leave the data slightly out of alignment. Depending on your needs for accuracy, this may be an entirely sensible choice. The amount of misalignment depends on the datums involved and the part of the world being mapped, but in the mismatch that North Americans usually encounter (between the World Geodetic System of 1984 and the North American Datum of 1983), it typically doesn't exceed a few feet. At most scales, the difference isn't noticeable.

The other option is to reconcile the systems through a geographic transformation. Transformations are often done in conjunction with a coordinate system projection. Like projections, they can be permanently applied to datasets with data processing tools, or they can be done on the fly in ArcGIS Pro. Transformations require some expert knowledge. There are default methods to convert one spheroid to another, but there aren't default fits, because the right fit depends on your area of interest. The table on "geographic (datum) transformations: well-known IDs, accuracies, and areas of use," at https://desktop.arcgis.com/en/arcmap/latest/map/projections/pdf/geographic_transformations.pdf, can help you find the right fit for an area of interest.

Lesson 4 Build the database

YOU'VE COME TO A TURNING POINT

in your project. In the early, exploratory phases, you looked at data—both in and out of maps—and evaluated it from various standpoints. You worked a lot with layer properties, but you didn't operate directly on the data. You'll do that now.

You have two main tasks in this lesson. Your first task will be to populate the LARiver_ParkSite project geodatabase with the datasets that you chose in lesson 3. Working with the data will be your introduction to the geographic data processing, or "geoprocessing," tools that make ArcGIS so powerful. That raises the question of whether the project really should have its own database. Couldn't you continue to use the collection of folders and files you've been working with so far? Yes, you could, but there are good reasons to consolidate the data and use the LARiver_ParkSite project geodatabase that was created automatically when you created the project:

- You can keep the project data in a single location, rather than having it distributed among various workspaces.
- You can keep the project data separate from datasets that you don't need.
- You can streamline the project data by eliminating unnecessary features and attributes while maintaining the original source data as a backup.
- You can impose uniform standards such as a common data format and consistent names for datasets.
- It will be easier to share the project data with others and to reuse it yourself.

Your second task in this lesson will be to complete the preparation tasks in the data requirements table. Besides using geoprocessing tools, you'll do spatial queries, table calculations, and other operations.

For more information, see the sidebar "Project database considerations" in this lesson.

Lesson Four road map

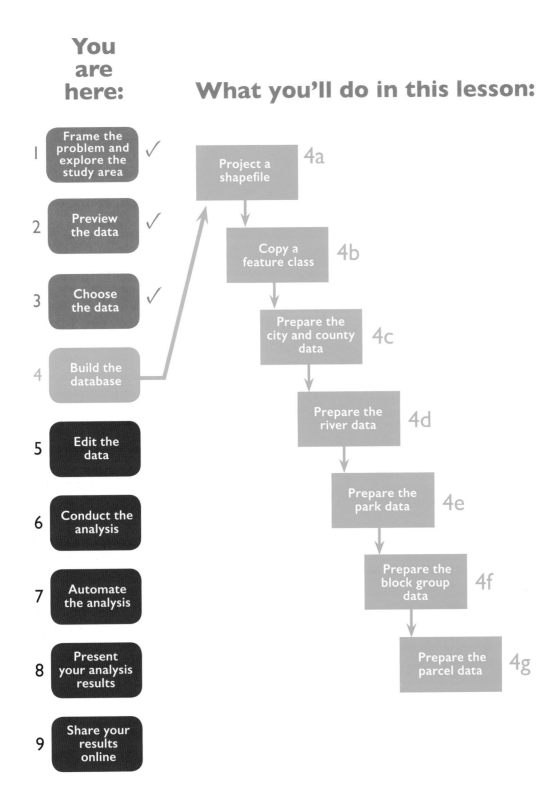

Project database considerations

For our simple database, we considered the following issues:

Data format: We put all the data into file geodatabase format. The geodatabase is the standard Esri spatial data format, and it has many advantages, large and small, over the shapefile format. We decided not to use feature datasets ⊞. Feature datasets are only required to support advanced data structures (topologies and networks) that aren't discussed in this book. They can also be used as thematic organizers (they're used this way in the Esri geodatabase in the source data for this project), but you won't have enough feature classes in your project database to make that worthwhile.

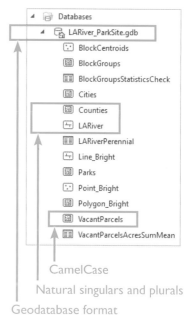

CamelCase
Natural singulars and plurals
Geodatabase format

Inputs and outputs: An analysis project has input and output data. You'll store your input, or starting, data in one geodatabase and your output data in a separate geodatabase. This isn't the only way to do it. You could store both the input and output data in the same geodatabase. The reason you're doing it this way is because eventually you might want to run variations on the analysis. This would lead to multiple sets of results, where each set could consist of several feature classes. Keeping track of all that data in one geodatabase would be confusing.

Naming conventions: Conventions aren't right or wrong in and of themselves. Ideally, they should be easy to remember and easy to apply, but the only real sign that a convention is "wrong" is that you don't stick to it. Our conventions are as follows:

- To separate word forms with capital letters, rather than spaces or underscores. For example, *VacantParcels* rather than *Vacant_Parcels* or *Vacant Parcels*. (This typographic convention is called "Pascal case" or "CamelCase.")
- To use natural singulars and plurals. For example, *LARiver* (one river) and *MajorRoads* (many roads).
- To use complete names, not abbreviations. For example, *MajorRoads*, not *MjrRds*.
- To prefer descriptive names to names that reflect software procedures. For example, *ExcludedParkAreas* is better than *Parks_Buffer_Dissolve_Erase*.

Coordinate system: Feature classes in a geodatabase don't have to be in the same coordinate system. (Feature classes in the same feature dataset do have to be in the same system.) Because ArcGIS Pro projects data on the fly, you don't have to worry about mixing and matching datasets. One reason, however, to keep all the data in the same coordinate system is that you don't have to think about making sure the data frame is set to the appropriate system. Also, it's considered better practice to make spatial edits to data in its native coordinate system than when it's projected on the fly. You'll project all the data in your database to the State Plane California Zone 5 coordinate system.

Maintenance and documentation: You don't have to worry about updates, security, quality control, or other issues that apply to keeping a large database up and running. In this case, maintaining the database means keeping it uncluttered. In the course of an analysis project, it's common to create datasets that are an important link in a chain of processes but that have no particular value once the end of the chain is reached. If you don't remove this "intermediate" data from the geodatabase, it accumulates quickly, often making it hard to distinguish critical outputs and results from byproducts. (This is especially true for others who may want to look at or use your data, but it's true for you as well.) In addition to keeping the database trim, you also want to keep it reasonably well documented. Output datasets that you intend to keep should have updated item descriptions to reflect the contents and purpose of the data.

Types of geodatabases

A geodatabase can be one of three types:

A **file geodatabase**, with extension .gdb, is stored as a collection of system files in a folder. It has no total size limit. Each dataset within the geodatabase has a default size limit of one terabyte, which can be increased to 256 terabytes. Multiple users can concurrently edit different datasets within the geodatabase.

A **personal geodatabase***, with extension .mdb, is a Microsoft® Access® database. It has a total size limit of two gigabytes. Only one user at a time can edit the database. File and personal geodatabases serve the same purposes; the personal geodatabase is just an older version of the technology.

An **ArcSDE® enterprise geodatabase** is stored in a relational database system such as Oracle®, Microsoft SQL Server™, PostgreSQL, or IBM® DB2®. (SDE stands for Spatial Database Engine.) Its size limits are determined by the database system in which it resides. Enterprise geodatabases support concurrent editing on individual datasets and provide database administration tools.

*Personal geodatabases are accessible with ArcMap but not ArcGIS Pro.

Exercise 4a: Project a shapefile

Your first task is to populate the geodatabase with starting data for the analysis. As needed, you'll convert the data format from shapefile to geodatabase and project it to the state plane coordinate system.

Open the data requirements table

Having this table open may help you keep the various preparation tasks in mind. As you work through the lesson, you can refer to it or not as you like.

1) **Using Windows Explorer, open the data requirements table from the MapsAndMore folder. If your table isn't up to date, download the results for lesson 3 from the book's resource web page, at esri.com/Understanding-GIS-3.**

Most of the datasets need some preparation. Actually, if you count format conversion and projection as preparation tasks, they all do. For now, start with a dataset that doesn't have anything listed in the Preparation column. The block centroids dataset in row 11 is one.

2) **Minimize the data requirements table.**

Get started

1) Start ArcGIS Pro and open your LARiver_ParkSite project.

2) Insert a new map and name it **Lesson4a**. As an alternative to using the Insert tab, try right-clicking the Maps folder in the Project pane and clicking New Map.

Open the Project tool

First of all, check the coordinate system of the Block Centroids layer to make sure it must be projected.

1) Open the Lesson2 map. (You can keep the Lesson4a map open as well.)

2) In the Contents pane, right-click block_centroids and click Properties.

3) In the Layer Properties dialog box, click the Source tab (on the left).

4) Scroll to the bottom of the dialog box and expand Spatial Reference.

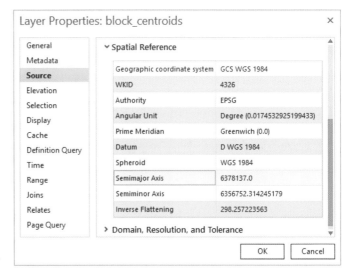

The geographic coordinate system is GCS WGS 1984. You'll project it to the state plane coordinate system. This isn't an on-the-fly projection of a layer in a data frame, but a geoprocessing operation on the feature class itself that converts all the WGS 1984 angular coordinates to state plane linear coordinates. You need a geoprocessing tool to do it.

5) Close the Layer Properties dialog box.

6) Close the Lesson2 map.

7) If necessary, return to the Lesson4a map by clicking the tab above the map.

8) On the Analysis tab, in the Geoprocessing group, click the Tools button to display the Geoprocessing pane.

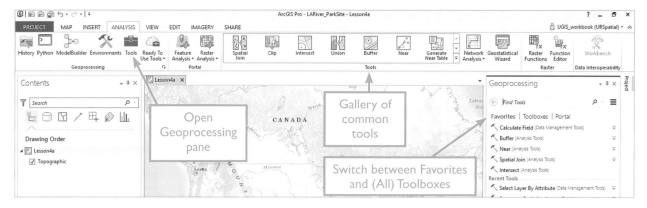

The Analysis tab and Geoprocessing pane allow you to access literally hundreds of tools for any number of tasks. The most common tools are displayed in a gallery across the top of the ribbon. This gallery can be customized to include the tools that make the most sense for your workflow. Also, notice that the Geoprocessing pane initially displays Favorites (tab at the top of the pane). This tab is also customizable.

9) **In the Geoprocessing pane, click Toolboxes.**

This pane displays all the tools included with ArcGIS Pro, organized logically into toolboxes. There are a lot of them, but most are specialized. The ones most useful to your project (and to much of the typical GIS work) are the Analysis, Conversion, and Data Management toolboxes.

10) **Expand Data Management Tools. Then expand the Projections and Transformations toolset and click the Project tool. Clicking opens the Project tool in the Geoprocessing pane.**

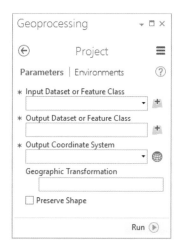

132   Lesson 4: Build the database

Most tools look basically the same as this one: a dialog box with several settings, called *parameters*. Required parameters, such as the input and output datasets or feature classes and the output coordinate system, are marked with a red asterisk.

The ArcGIS Help panel summarizes the tool's function. This tool takes an input dataset and creates a new output dataset in whatever coordinate system you choose.

11) **In the upper-right corner of the pane, click the Help button ⓘ. This button opens a detailed help page specific to this tool and its usage. Close the web browser when you are finished looking at the help page.**

Project the block_centroids shapefile

Now you'll fill out the parameters and run the tool. The first thing to specify is the input dataset, the block_centroids shapefile.

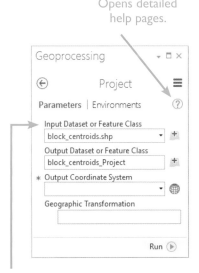

Opens detailed help pages.

Hover or click for parameter information.

1) **In the Project tool, hover over the red asterisk, which turns into an information button ⓘ, next to the Input Dataset or Feature Class parameter. It reads, "The feature class, feature layer, or feature dataset to be projected."**

These parameter information windows help decipher the parameter names, which can often be rather technical or cryptic.

2) **Click the browse button ⊞ on the right of the parameter. Clicking opens a browse dialog box. Click the Folders item on the left of the dialog box, and then browse to ParkSite > SourceData > census. Double-click block_centroids to add it as a parameter to the tool.**

Once you drop the input dataset in place, ArcGIS Pro can fill out some of the other parameters for you. It reads the input coordinate system (GCS_WGS_1984) from the metadata and directs the output to the default geodatabase. This geodatabase is where you want the data to go, but it's not the feature class name you want.

3) **In the Output Dataset or Feature Class box, highlight the entire path and press the Delete key. In its place, type BlockCentroids. The output path to the LARiver_ParkSite project geodatabase is restored.**

4) **Next to the Output Coordinate System parameter, click the Select coordinate system button ⊕.**

5) **Expand the Projected Coordinate Systems folder, then the State Plane folder, and then the NAD 1983 (US Feet) folder.**

Exercise 4a: Project a shapefile   133

▶ Be careful to use the NAD 1983 folder with that exact name. Several other versions of the name look similar, such as NAD 1983 (2011) (US Feet).

6) Scroll down and click **NAD 1983 StatePlane California V FIPS 0405 (US Feet)** and then click **OK**.

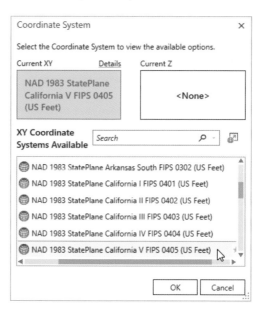

Because the input and output datasets have different geographic coordinate systems—WGS 1984 and NAD 1983, respectively—a geographic transformation has been listed automatically. When you project layers on the fly, you can bypass geographic transformations, but the Project tool makes you do them.

The default transformation chosen by ArcGIS Pro is usually the best choice—in this case, WGS_1984_ (ITRF00) _To_NAD_1983. There may be times when you want to use a different one; for example, to conform to standard practice in your organization. In this case, you'll keep the default instead of changing the transformation, but be aware that you can change it if necessary. For more information about geographic transformations, see the sidebar "Datums" in lesson 3.

7) Check your Project tool settings against the figure and click **Run** ▶ at the bottom of the Geoprocessing pane.

When the process is completed, the new BlockCentroids layer is added to the map. The BlockCentroids feature class with the coordinate system you specified has also been saved to the LARiver_ParkSite project geodatabase.

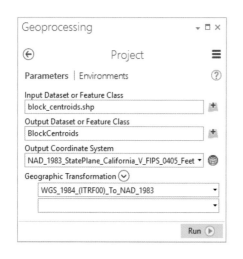

134   Lesson 4: Build the database

8) Close the Geoprocessing pane by pressing X in the upper-right corner of the pane.

9) If necessary, zoom to the BlockCentroids layer.

10) In the Project pane, expand the Databases folder, if necessary, and expand the project geodatabase. Confirm that the new BlockCentroids is present.

▶ If you don't see the feature class listed, right-click the database in the Project pane and click Refresh (or press F5 on the keyboard).

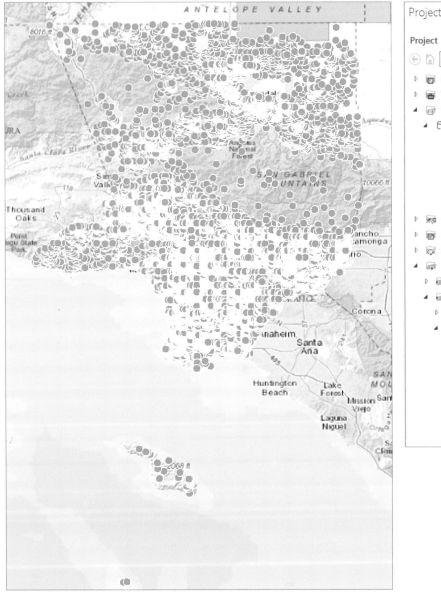
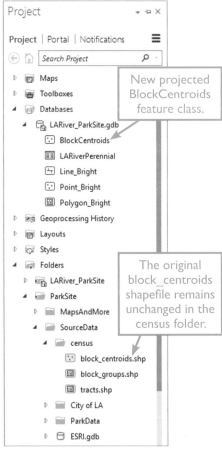

Exercise 4a: Project a shapefile 135

11) Confirm that the coordinate system of BlockCentroids is State Plane California Zone V and that its geographic coordinate system is GCS_North_American_1983.

▶ How? Open the BlockCentroids layer properties under Source > Spatial Reference.

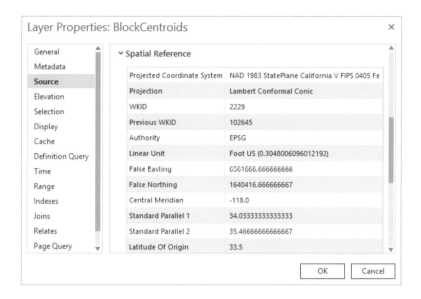

12) Open the Map Properties of the Lesson4a map and set State Plane California Zone V as a Favorite coordinate system by clicking Add to Favorites on the right of the projection.

▶ How? In the Contents pane, right-click the name of the map and click Properties.

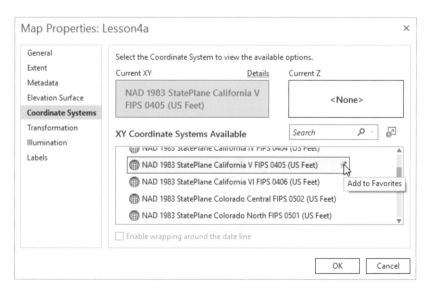

Setting a favorite projection saves the projection to the favorites group so it is easier to find.

13) Click OK on the Map Properties dialog box.

14) On the Analysis tab, click History.

The Project pane opens on the right showing the geoprocessing history. All geoprocessing operations are logged here in a list displaying which tools were run at what date and time along with any success or error messages. Hover over Project to see more details about tool parameters. Results are especially useful when a process fails because the error is reported in the Messages section.

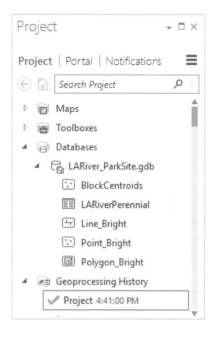

You can reopen and rerun a tool with all the parameters already set from this list. Simply double-click or right-click the tool and click Open to open the tool. This is a convenient way to rerun the tool if you want to simply change a parameter or need to rerun the tool for any reason.

15) Reduce the Geoprocessing History item in the Project pane.

In the next exercise, you'll add a feature class to the LARiver_ParkSite project geodatabase using the Copy tool.

Exercise 4b: Copy a feature class

In this exercise, you'll copy the Parkland feature class to the project geodatabase.

Get started

1) If necessary, start ArcGIS Pro and open your LARiver_ParkSite project.

You'll learn a useful skill of setting the geoprocessing environment. Environment settings are default values or conditions applied automatically to all geoprocessing operations.

2) Make a copy of the Lesson4a map and rename it **Lesson4b**.

Set the geoprocessing settings

There are many environment settings, but you'll concentrate on the output coordinate system.

1) On the Analysis tab, click Environments, which opens the Environments pane.

2) In the Output Coordinate System drop-down list, click BlockCentroids. This will apply the coordinate system from BlockCentroids that you defined in exercise 4a.

3) Click OK.

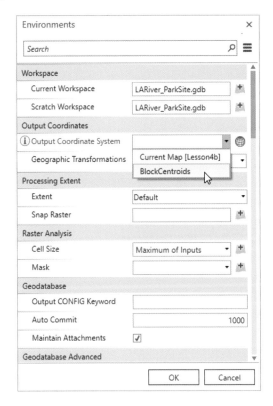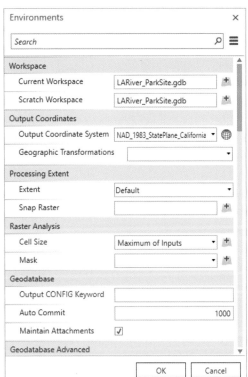

Copy Parkland

The Parkland feature class is in the WGS 1984 geographic coordinate system and must be projected.

1) On the Analysis tab, click the Tools button to display the Geoprocessing pane.

2) In the Geoprocessing pane Search box above the tools list, type copy. As you type, ArcGIS Pro presents a list of possible matches.

3) Click the Copy Features tool.

4) Click the browse button on the right of the Input Features parameter. This opens a browse dialog box. Click the Folders item on the left of the dialog box, and then browse to the ParkSite > SourceData folder. Double-click ESRI.gdb to open, and then double-click Landmark to expand.

5) Click the Parkland layer.

6) Press Enter.

Exercise 4b: Copy a feature class   139

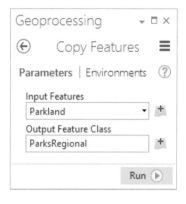

7) For Output Feature Class, type ParksRegional. Confirm that the feature class is saved to the LA River geodatabase.

8) Check your Project tool settings against the figure. Click Run at the bottom of the Geoprocessing pane.

When the process is completed, the new ParksRegional layer is added to the map. The ParksRegional feature class with the coordinate system you specified has also been saved to the LARiver_ParkSite project geodatabase. Thanks to the environment setting, the ParksRegional feature class was also projected to the state plane coordinate system. As long as that setting remains in place—which it will, within the current project, until you change it—any geoprocessing tool you run will also cause the input dataset to be projected. As usual with geoprocessing operations, the outputs are new feature classes, and the original inputs are unchanged.

Inspect the results

Now you can see what you've accomplished.

1) If necessary, add the ParksRegional layer to the Lesson4b map, and confirm that the coordinate system is in state plane by opening the Layer Properties > Source > Spatial Reference of the layer.

2) In the Project pane, expand the LARiver_ParkSite geodatabase, if necessary.

It should contain two feature classes: BlockCentroids and ParksRegional, along with the three _Bright layers. It should also include one table: LARiverPerennial.

3) Save your project by clicking the Save button.

If you're continuing to the next exercise, leave ArcGIS Pro open; otherwise, close ArcGIS Pro, and close the data requirements table.

Exercise 4c: Prepare the city and county data

Now you'll move on to data that needs more preparation. In the data requirements table, you listed City_ply and County as data needed for the final map. You made notes to reduce their extent because both feature classes cover the entire United States.

Processing data to get rid of extra features or attributes is a matter of choice. You could just as well leave the datasets intact and instead apply appropriate definition queries to the City_ply and County layers in the final map. One point in favor of small datasets is that they generally display and process faster. Likewise, an attribute table is easier to read if it's not cluttered with unnecessary attributes. If you later decide to share the project data with others, it will also be convenient to have a smaller package.

You must be careful in deciding which features and attributes are "unnecessary," because once you get rid of them, they're gone—gone, at least, from that dataset. You'll still have the original source data as a backup, so you can recover from mistakes. You probably wouldn't delete features or attributes that weren't preserved in another feature class.

Add a new map and change the coordinate system

1) **If necessary, start ArcGIS Pro and open the LARiver_ParkSite project.**

2) **Insert a new map and name it Lesson4c.**

3) **Change the coordinate system of the new map to the state plane system that you've been using for this project.**

▶ How? In the Contents pane, right-click the map and open the Properties dialog box. Go to Coordinate Systems, and then navigate to the state plane system. If you added it to your favorites, it should be listed in the Favorites group.

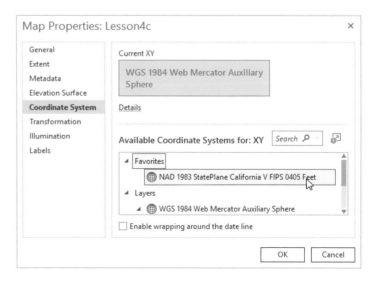

Add data to the map

You'll add the City_ply and County feature classes as layers to the map. Then you'll select the features you want to keep and copy the selections to new feature classes.

1) In the Project pane, in the SourceData folder, under ESRI.gdb, expand the Boundary feature dataset.

2) Press and hold the Ctrl key and click the City_ply and County feature classes to highlight both.

3) Drag the datasets to the map window. The two layers are added to the map.

4) In the Contents pane, click the Lesson4c header to clear the selected layers.

5) If necessary, drag City_ply above County.

6) Change the symbols to colors that suit you.

7) Zoom to Southern California.

Select counties

In this section, you'll select the county features that you want to keep.

1) Turn off the City_ply layer.

2) In the Contents pane, right-click County and click Label.

The counties are labeled by name. Label positions change dynamically with the map scale and extent, so yours may be placed differently.

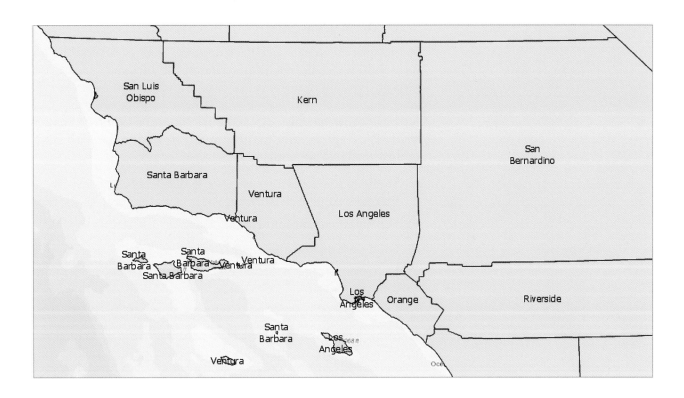

3) If necessary, pan the map to put Los Angeles County roughly in the middle of the display.

4) On the Map tab, click the Select tool. Click anywhere inside Los Angeles County to select it.

The selected feature is outlined in bright blue. This feature is probably the only county feature you'll need for the final map, but to be safe, you'll select its neighbors as well. You'll do this using a spatial query.

Exercise 4c: Prepare the city and county data

5) **On the Map tab, click Select By Location.**

Spatial queries evaluate spatial relationships among features. These relationships include proximity, containment, adjacency, and intersection. Typically, features in one layer (the target layer) are selected according to their relationship to features in another layer (the source layer). Sometimes, as you'll work with here, the target and source layers are the same so that features are selected on the basis of their spatial relationship to the selected feature(s).

6) **Confirm that County is selected as Input Feature Layer (it may already be selected).**

7) **For Relationship, confirm that Intersect is selected.**

8) **For Selecting Features, click County.**

9) **In the Selection type drop-down list, click Add to the current selection. You want to keep your currently selected feature, Los Angeles County, and select more features.**

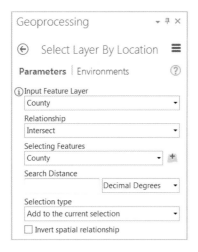

Touching features are considered "intersecting," so in plain English, this query means that counties bordering Los Angeles County will be selected and added to the currently selected Los Angeles County.

10) **Compare your dialog box to the figure and click Run.**

On the map, Los Angeles, Ventura, Kern, San Bernardino, and Orange Counties are selected.

Hide unnecessary fields

The County layer will serve a purely cartographic purpose. That means the only attributes you'll need are those you might want for labeling or symbology.

1) **Under Contents, right-click the County layer and click Attribute Table.**

2) **Scroll quickly across the attribute columns to see what attribute information is provided.**

There are about demographic and economic attributes— none of which you need. Removing them will tailor your data to its cartographic purpose.

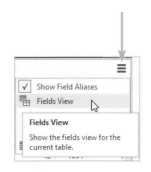

3) **Click the Menu button ≡ in the upper-right corner of the attribute table and click Fields View. (Alternatively, you can access this view by clicking the Fields button on the Data ribbon.)**

The fields view appears below the map, where you can customize how attribute fields are displayed on the map. The scrolling box on the left lists the fields in the table. Each field's check box can be unchecked to hide the field. Field visibility is a layer property that you can turn on and off as you like—a hidden field isn't deleted. Nevertheless, the setting affects the way data is copied. When you copy the feature class, hidden fields will be left out.

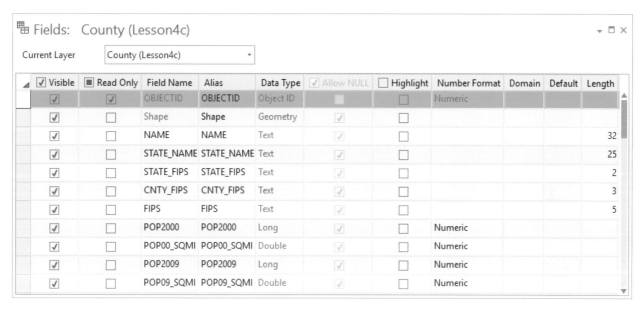

4) Above the list of fields, click to clear the Visible check box to turn all fields "off" (meaning they are hidden in the layer).

5) Click the boxes in the Visible column next to NAME and STATE_NAME to turn just these two fields back on.

6) On the Fields tab, click the Save button.

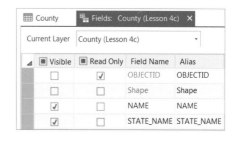

7) Close the Fields View pane (open pane below the map).

8) If necessary, open the County attribute table to see that only the two fields "marked" visible are displayed.

9) Close the attribute table and save your project.

Exercise 4c: Prepare the city and county data   145

Copy the selected features to a new feature class

Using the Copy Features tool again, you'll copy the five selected features and their visible attributes to a new feature class.

1) On the Analysis tab, in the Geoprocessing group, click History.

2) The Project pane opens on the right showing the geoprocessing history.

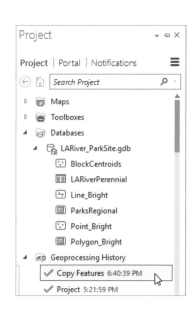

3) Double-click the Copy Features tool to open it again.

4) In the Copy Features tool dialog box, click the Input Features drop-down arrow and click County.

You can choose input features from a drop-down list when there are layers in the map.

5) In the Output Feature Class box, highlight and delete the entire path. Type Counties.

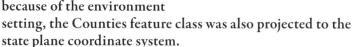

6) Click Run. When the tool is finished running, a Counties feature class is added to the LARiver_ParkSite geodatabase, and a corresponding layer is added to the map. As before, because of the environment setting, the Counties feature class was also projected to the state plane coordinate system.

7) Open the attribute table of Counties to confirm that the only fields copied from the County layer are NAME and STATE_NAME. The other four (OBJECTID, Shape, Shape_Length, and Shape_Area) are created and updated automatically by ArcGIS Pro.

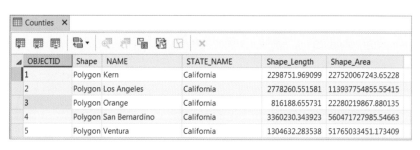

146   Lesson 4: Build the database

The Counties layer consists of the five counties you selected. The islands are part of the Los Angeles and Ventura Counties features. Features with discontinuous geometry (islands are a common example) are called "multipart" features. If you click on an island with the Explore tool on the Map tab, you'll see the entire county flash blue.

8) **Open the Layer properties of Counties and confirm that the Spatial Reference under Source is NAD 1983 StatePlane California V FIPS 0405 Feet.**

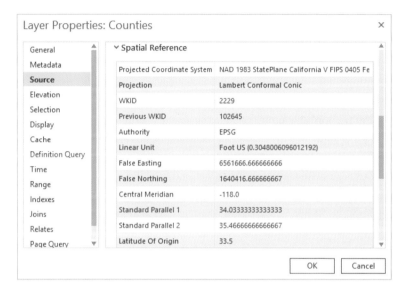

9) **Close the Counties Layer Properties dialog box and attribute table.**

10) **Under Contents, remove the (old) County layer to avoid any confusion.**

▶ How? In the Contents pane, right-click County and click Remove.

11) **Right-click the (new) Counties layer and click Zoom To Layer.**

Select cities

As with the counties, you don't know exactly which cities you need for the final map. You'll make an interactive selection that takes in an indefinite, but ample, number of features.

1) **In the Contents pane, turn on the City_ply layer.**

2) If necessary, drag City_ply above Counties.

3) At the top of the Contents pane, click the List by Selection tab, and then click to clear the Counties layer. This step makes City_ply the only selectable layer (so when you select by dragging to draw a box, the only features selected will be from that layer).

4) On the Map tab, in the Selection group, click the Select Features tool.

5) Use the mouse to draw a selection box similar to the one in the figure (your box can be bigger or smaller).

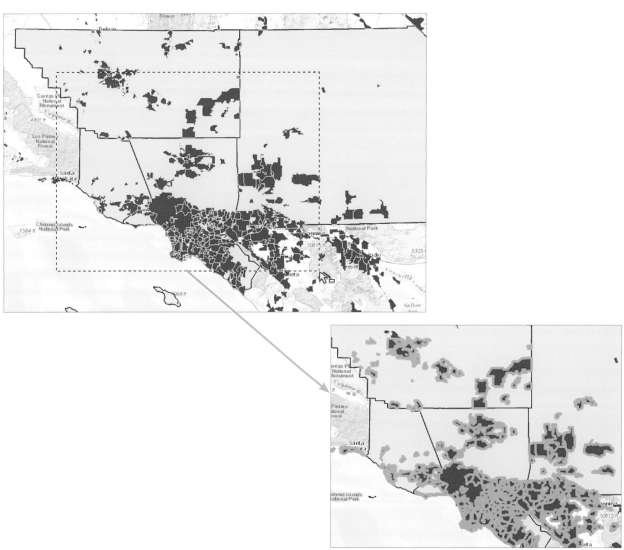

When you release the mouse button, cities that are entirely or partially within the box are selected.

Note that the List By Selection tab on the Contents pane shows the number of selected features next to the layer name. In the example, 331 cities have been selected. It doesn't matter if you selected more or fewer features.

6) At the top of the Contents pane, switch back to the List By Drawing Order tab.

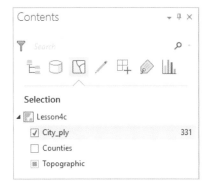

Turn off fields to prepare for export

As you did with the counties, you'll get rid of attributes you don't need for your analysis.

1) Open the attribute table of the City_ply layer, and use the Menu button to open the fields view.

2) Click to clear the Visible check box at the top of the table to turn all fields "off."

3) Under the Visible column, click the check boxes next to NAME and ST (an abbreviation for State) to turn these two fields back on. These are the only attribute fields you need for your project.

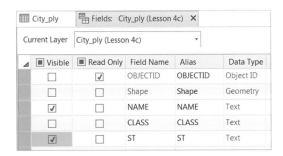

4) Compare to the figure in step 3, and then click Save on the Fields tab.

5) Close the fields view (the open pane below the map).

Copy the selected features to a new feature class

Using the Copy Features tool, you'll copy the selected features and their visible attributes to a new feature class. Instead of using the geoprocessing history, you'll use another method to open the Copy Features tool.

1) In the Contents pane, right-click the City_ply layer and click Data > Export Features. This option displays the same Copy Features tool in the Geoprocessing pane that you worked with earlier.

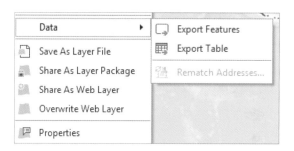

2) In the Output Feature Class parameter of the Copy Features tool, highlight and delete the entire path. Then type Cities.

3) Compare your tool to the figure and click Run.

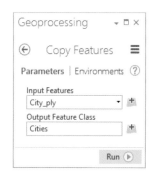

When the tool is finished running, a Cities feature class that contains just those features you have selected, plus the visible fields you selected, is added to the LARiver_ParkSite geodatabase. A corresponding layer is added to the map in the coordinate system you have chosen for the project.

4) In the Contents pane, remove the (old) City_ply layer.

5) Open the attribute table of the (new) Cities layer to verify that the export worked as expected.

The table contains the two attributes you kept, plus the software-managed attributes.

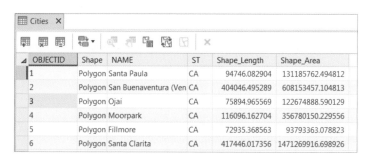

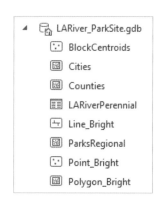

6) Close the table and any other open tables or panes under or on top of the map view.

7) If necessary, in the Project pane, expand the LARiver_ParkSite geodatabase. It should now contain four feature classes: BlockCentroids, Cities, Counties, and ParksRegional, along with the three _Bright layers and the LARiverPerennial table.

8) Save your project.

9) If you're continuing to the next exercise, leave ArcGIS Pro open; otherwise, close ArcGIS Pro.

Exercise 4d: Prepare the river data

In row 5 of the data requirements table, you made a note to dissolve the LARiver dataset. This action will combine the river's 265 individual line segments into a one-line feature, simplifying the task of creating a buffer around the river in lesson 6. The Dissolve tool creates a new feature class. When you run the tool, you'll convert the LARiver shapefile into a geodatabase feature class as part of the operation.

Dissolve the river

A few common geoprocessing tools, including Dissolve, can be accessed from the gallery on the Analysis tab.

1) If necessary, start ArcGIS Pro and open the LARiver_ParkSite project from your recent projects.

2) Make a copy of the Lesson4c map and rename it Lesson4d.

3) Open map Lesson4d.

4) On the Analysis tab, expand the Tools gallery by clicking the drop-down arrow on the lower right. The Tools gallery provides quick access to some of the most commonly used tools among the hundreds included with ArcGIS.

5) Scroll down near the bottom to find the Dissolve tool in the Manage Data group. Click the tool to open it.

6) For the Input Features parameter, click the browse button and browse to the LA River shapefile at UGIS\ParkSite\SourceData\City of LA\LARiver.shp.

7) **Change the Output Feature Class name from LARiver_ Dissolve to LARiver.** Outputs of tools default to your project's "home" geodatabase. You'll notice that the entire path to the geodatabase will be displayed when you hover your pointer over the text box. However, you only need to change the name of the new feature class (the last part of the path).

8) **In the Dissolve_Field(s) area, click to select the NAME check box.**

Selecting this check box preserves the NAME field, which you want to keep, in the output attribute table. It also affects the way the tool works. Instead of creating a single output feature, ArcGIS Pro will create a single feature for each unique value in the checked field. In this case, it results in the same thing because the NAME field has only one value ("Los Angeles River"). If you selected a different field, however, the output feature class would have more than one feature.

9) **Compare your tool to the figure and click Run.**

When the tool is finished running, a feature class is added to your project geodatabase, and a layer is added to the map.

10) **Open the attribute table of the LARiver layer. Note that the Dissolve tool took the shapefile with its 265 features down to one record.**

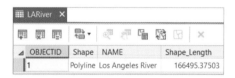

Except for the NAME field you selected, the attributes from the original table have been dropped for the one remaining record. Confirm that the length values match.

11) **Close the attribute table.**

12) **Symbolize the river in a shade of blue and increase its width.**

For more information, see the sidebar "Dissolving features" later in this lesson.

Exercise 4e: Prepare the park data

The ParksRegional feature class is already in your project geodatabase, but you still have the preparation task of loading data into it. That's because this dataset is missing the two park features that you saw in the NewParks shapefile.

Add data to the map

Now you can remind yourself which two parks we're talking about.

1) Make a copy of the Lesson4d map and rename it Lesson4e.
2) Open map Lesson4e.
3) From the project geodatabase (LARiver_ParkSite.gdb), add ParksRegional to the map. Optionally, symbolize the layer in a shade of green.
4) Under SourceData, expand the ParkData folder and add NewParks.shp. Optionally, symbolize this layer in a different shade of green.
5) Zoom to the NewParks layer.

Dissolving features

You can dissolve a feature class either by geometry or by an attribute. If you dissolve by geometry, common boundaries between features disintegrate. The number of output features depends on whether the "Create multipart features" box is checked. If it's checked (the default), you'll get one output feature with discontinuous geometry. If you clear the check box, you'll get a unique, single-part feature for each spatially distinct area. In either case, the output table has no attributes except for those maintained by ArcGIS Pro.

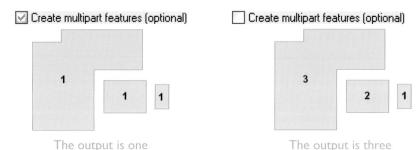

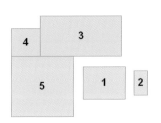

Five input features will be dissolved by geometry

The output is one multipart feature

The output is three single-part features

If you dissolve by an attribute, common boundaries between features disintegrate only if the features have the same attribute value. Again, the number of output features depends on whether multipart features are created. If so, all input features with the same attribute value will become one output feature (with discontinuous geometry). If not, you'll get a unique feature for each spatially distinct area. The attribute used for the dissolve is preserved in the output table.

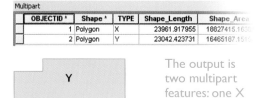

Five input features will be dissolved on a TYPE attribute

The output is two multipart features: one X feature and one Y feature

The output is four single-part features: two X features and two Y features

In addition to the attribute used for the dissolve, you can set a "statistics field" to summarize the values of dissolved features for another attribute. In this example, the features are dissolved by TYPE, and SUM_ACRES is a statistics field (based on an ACRES field in the original table) that calculates the total acreage of the dissolved features.

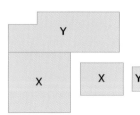

154   Lesson 4: Build the database

Compare attributes

You want to get these two NewParks features, along with their attributes, into the Parks feature class. You face a complication in that the datasets have different attributes.

1) **Open the ParksRegional attribute table.**

It has 1,209 records. Its user-managed attributes are NAME, TYPE, and ACRES.

The TYPE attribute describes whether the park is a local, state, or national park.

2) **Open the NewParks attribute table.**

Its user-managed attributes are NAME, CATEGORY, ADDRESS, TELEPHONE, and HOURS.

Both tables have a NAME field. CATEGORY is similar to the TYPE field in the ParksRegional table in that it identifies the parks as state parks. The other fields don't have equivalents in the ParksRegional table.

3) **Close the open attribute tables.**

Use Append to load data

The Append tool loads data from one or more datasets and appends the features into an existing dataset. It also provides tools for dealing with attribute mismatches.

1) **On the Analysis tab, find the Append tool in the Tools gallery. It's right next to the Dissolve tool you used in exercise 4d.**

2) **For Input Datasets, click NewParks. This dataset is the one to be loaded. Note that you can append multiple datasets, but you will be using only one for this exercise.**

3) **For Target Dataset, click ParksRegional. This dataset is the one in which the data will be loaded.**

4) **Change Schema Type to Use the Field Map to reconcile schema differences.**

This schema type displays the Field Map parameter with additional options to describe how attribute values will be loaded from the input to the target dataset (this is also referred to as "mapping fields"). The Output Fields column on the left side lists the available attribute columns in the target dataset. The Source column on

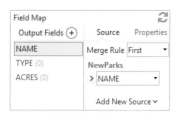

the right side lets you pick which field(s) you want from the input dataset(s). If the field names match, ArcGIS Pro assumes that the fields should be matched.

5) **In the Field Map table, click NAME in the Output Fields column. Notice that NAME has a match because there is a field named NAME in the NewParks shapefile.**

Next, you will "map" the CATEGORY field from NewParks to the TYPE field in Parks.

6) **In the list of Output Fields, click TYPE.**

7) **Click Add New Source, and then in the list of fields, click CATEGORY.**

8) **At the bottom of the list, click Add Selected.**

You have set up the field map so the values from the NAME field in NewParks will be loaded into the NAME field in ParksRegional. The values from the CATEGORY field in NewParks will be loaded into the TYPE field of ParksRegional.

There is no ACRES field (by this or any other name) in the NewParks table, so there is no need to map this field.

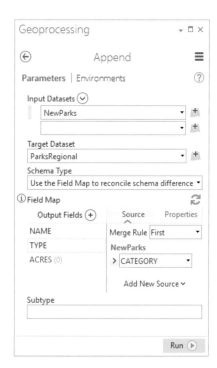

The address, telephone, and hours information in NewParks won't be loaded because ParksRegional has no appropriate fields to accept the values. If these fields were necessary, you would have to add these fields to the target dataset before appending it.

9) **Compare your tool to the figure and click Run.**

Review the results

1) **Turn off the NewParks layer in the map and notice that the two features have been added to ParksRegional.**

2) **Open the ParksRegional attribute table and confirm that the two new park records were added at the bottom of the table. You should now have a total of 1,211 records in the table.**

3) **Close the attribute table.**

For more information, see the sidebar "Field data types" later in this lesson.

Make a spatial selection in the ParksRegional layer

It's not one of your preparation tasks to reduce the Parks data, but the spatial extent of this feature class is bigger than it needs to be. Getting rid of extra features may help speed up some analysis operations in lessons 6 and 7. In the next steps, you'll use Select By Location to get a copy of just those parks that are within three miles of the LA River.

1) Zoom to the ParksRegional layer.

2) Turn off the Cities layer.

Those big national forests to the north are definitely outside your area of interest.

3) In the Contents pane, select the ParksRegional layer, and on the Map tab, in the Selection group, open the Select By Location tool.

4) If necessary, click ParksRegional in the Input Feature Layer drop-down list (this layer should already be selected for you if you selected the layer in the Contents pane).

5) In the Relationship drop-down list, click Within a distance geodesic. For the Selecting Features parameter, click LARiver.

6) Use the Search Distance parameter to define 3 Miles as the distance.

7) Leave the Selection type parameter as New selection.

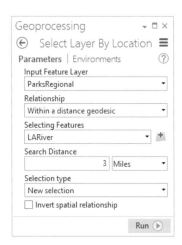

The spatial query will select parks within three miles of the river. By using the geodesic option, ArcGIS Pro will use the shortest path on the surface of a globe rather than the planar distance of the map projection, which is the shortest path in a 2D Cartesian plane. If any bit of a park is within the geodesic distance, the feature will be selected. All the unselected features will be well out of your area of interest.

8) Compare your tool to the figure and click Run.

9) Zoom to the City of Los Angeles bookmark.

Field data types

One of the essential properties of a field in an attribute table is its data type, which defines the type of information the field can hold. When a new field is added to a table, its data type is specified and can't be changed thereafter. There are several field data types, but those used most often are text (sometimes called "string") and the four numeric types: short integer, long integer, float, and double.

Text fields are for descriptions, codes, noncomputational numbers (such as postal codes or telephone numbers), and the like. The default length of a text field is 255 characters. If you know that your entries will be shorter—for example, if you're storing two-letter state abbreviations—you should set the length property accordingly. In the interest of saving space, you can also use codes in place of long descriptions. In a geodatabase, codes and their associated values can be managed through a so-called "attribute domain" (not discussed in this book).

For numeric data, use short or long integers when the values are whole numbers, such as population or number of items sold. Whether to use the short or long integer type depends on how large the numbers are (see the associated table). Use floats or doubles when the values are fractional. Floats store six digits with precision (for example, 12345.6 or 1.23456). Numbers with more digits are stored with some rounding, which may slightly affect calculations. Doubles are stored with 15 digits of precision. Floats are suitable for most statistical calculations, such as population density and average income. Doubles are recommended for fields storing geographic measurements and calculations. ArcGIS Pro uses the "double" type in its Shape_Length and Shape_Area fields.

Table 4-1 summarizes the field data types available for geodatabase feature classes. Not all data types are available for shapefiles.

Table 4-1. Field data types for geodatabase feature classes	
Text	Letters, numbers, special characters
Short integer	Whole numbers from −32,768 to 32,767 (uses 2 bytes)
Long integer	Whole numbers from −2,147,483,648 to 2,147,483,647 (uses 4 bytes)
Float	Fractional numbers: precise to 6 digits, then rounded (uses 4 bytes)
Double	Fractional numbers: precise to 15 digits, then rounded (uses 8 bytes)
Date	Dates and/or times
BLOB	Binary large objects, including special feature types such as geodatabase annotation
Raster	Small images
GUID	Unique feature/record identifier for features in distributed geodatabases
ObjectID	Unique feature/record identifer
Shape	Feature geometry type (for example, point, line, or polygon)

On the map, park features within three miles of the river are selected.

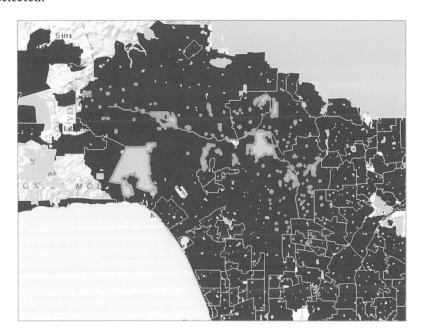

10) At the top of the Contents pane, click the List by Selection button.

11) Look at the layer list and note how many parks were selected. Again, these are the parks that are within three miles of the LA River. Confirm that you have 169 selected records.

12) Switch back to the List By Drawing Order tab.

Copy features

Now you'll copy the selected parks in the ParksRegional layer to a new feature class named Parks.

1) Open the toolbox by clicking the Tools button on the Analysis tab.

2) Note that Copy Features is listed as one of your Favorites. Click Copy Features to open the tool. If necessary, search for it in the Search box.

3) For Input Features, click ParksRegional.

4) Name the output Parks.

5) Compare your tool to the figure and click Run.

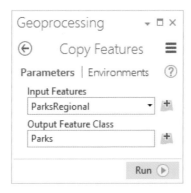

When the tool is finished running, Parks is created in your project geodatabase, and the layer is added to the map.

6) Under Contents, turn off the ParksRegional layer.

7) Resymbolize Parks in any shade of green.

8) Zoom to the Parks layer and open its attribute table. There should be 169 records to match the parks you selected previously.

9) Close the table.

You don't need the ParksRegional feature class anymore so you'll delete it from the geodatabase.

10) Make sure ParksRegional—not Parks—is highlighted in the Project pane. In the Project pane window, under your project geodatabase, right-click ParksRegional and click Delete.

The feature class is deleted from your project geodatabase, and the corresponding layer is removed from the map. (You still have the original Parkland data in your SourceData folder, so you can recover from mistakes if you have to.)

▶ If the ParksRegional layer is not removed from the map automatically, right-click it and click Remove.

11) Save your project.

Exercise 4f: Prepare the block group data

In rows 8 and 9 of the data requirements table, you identified some block group attributes—population density and age under 18—that you must calculate from your existing data. You'll add new fields to the table, and then calculate values into them.

Add data to the map

First, you'll add the block_groups shapefile to the map and have another look at it before you copy it to the project geodatabase.

1) If necessary, open the LARiver_ParkSite project in ArcGIS Pro.
2) Make a copy of the Lesson4e map and rename it Lesson4f.
3) Open the Lesson4f map.
4) In the Project pane, under SourceData, expand the census folder. Drag block_groups.shp to the map.
5) Zoom to the block_groups layer.

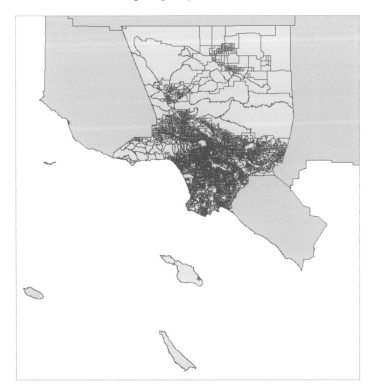

6) Open the block_groups attribute table and scroll across the attributes.

You need only a handful of the two dozen or so attributes in the table. As with the parks, you didn't make a note to reduce this dataset, but it will be easier to work with the important attributes if you don't have to keep scrolling past the unimportant ones.

Turn off unnecessary fields

You'll turn off fields so they won't be included unnecessarily when you copy the dataset. You'll turn off the fields you don't need.

1) Click the Menu button in the upper-right corner of the attribute table and click Fields View. (Alternatively, you can access this view by clicking the Fields button on the Data ribbon.)

2) Above the list of fields, click to clear the Visible check box to turn all fields off.

3) Click the check boxes for the following fields to turn them back on:

   - FID (a unique user-managed identifier)
   - TOTPOP_CY (total population)
   - POP18UP_CY (population over 18)
   - MEDHINC_CY (median household income)

4) Change the TOTPOP_CY alias to Total Pop.

5) Change the POP18UP_CY alias to Pop Over 18.

6) Change the MEDHINC_CY alias to Median HH Income (HH is short for Household).

7) Compare the fields view to the figure.

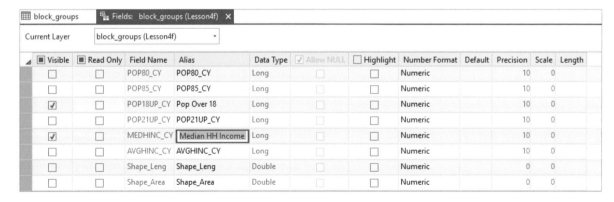

8) On the Fields tab, in the Changes group, click Save.

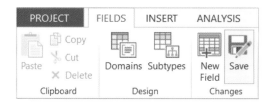

9) Close the Fields View pane and any open attribute tables.

Export block groups to the project geodatabase

Now you'll use the Copy Features tool to copy and project the block group feature class into your project geodatabase.

1) Right-click block_groups and click Data > Export Features.

2) In the Output Feature Class box, type BlockGroups to match your naming convention.

3) Compare your tool to the figure and run the tool.

4) Close the Geoprocessing pane.

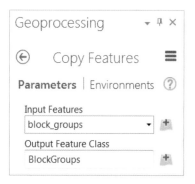

When the tool is finished running, a new feature class is created in the project geodatabase, and a layer is added to the map.

5) Under Contents, remove the (old) block_groups layer.

6) Open the attribute table of the (new) BlockGroups layer and verify that the changes were applied as expected.

Calculate population density

Population density is the ratio of population to area. The Shape_Area field stores the area of each block group on the basis of the units of the coordinate system. (This field was added automatically by ArcGIS Pro when you converted the block groups to a geodatabase format.) You know the units are square feet because those are the measurement units specified by your coordinate system. Since you must express density in terms of people per square mile, you're going to add two fields to the BlockGroups attribute table: one to store area in square miles and another to store population density expressed in square miles. Then you'll make the calculations for both fields.

1) At the top of the BlockGroups attribute table, click the Add Field button 🗔. This button opens the Fields pane with a new row started at the bottom.

2) For Field Name, type SQMILES.

3) For Alias, type Sq Miles.

4) For Data Type, click in the cell and then click Double in the drop-down menu.

5) On the Fields tab at the top of the application, click New Field 🗔.

6) Add another field and give it the following properties:
    - Name: POPDENSITY
    - Alias: Pop Density
    - Type: Float

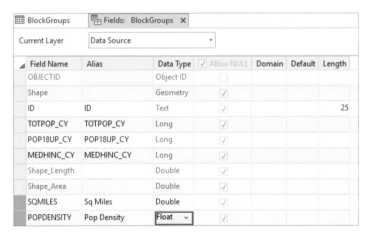

7) Compare your Fields pane to the figure in step 6 and click Save on the Fields tab.

8) Close the Fields pane and return to the BlockGroups attribute table. Note that the two new fields now appear in the attribute table.

9) In the BlockGroups attribute table, right-click the Sq Miles field heading and click Calculate Field. Alternatively, click the Sq Miles column header, and then click the Calculate Field 🗔 button at the top of the attribute table.

The Geoprocessing pane opens with the Calculate Field tool displayed. This tool allows you to populate attribute values using a value, mathematical expression, or simple code written in the Python scripting language. You will be using a simple Python expression to calculate the Sq Miles field, with each block group's area in square miles.

10) Confirm that BlockGroups is set as the Input Table parameter and Sq Miles is set as the Field Name parameter.

11) In the text box under SQMILES =, enter the following text:

    `!shape.area@squaremiles!`

12) Compare to the figure in step 11 and click Run.

ArcGIS Pro calculates the area in square miles for each record and adds the values to the table.

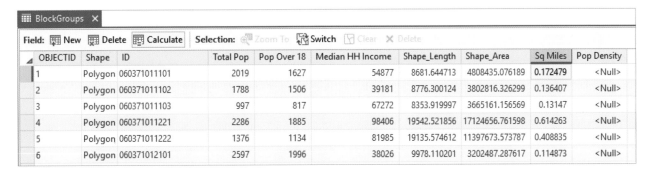

Exercise 4f: Prepare the block group data 165

Calculate another field

1) In the table, right-click the Pop Density field heading and click Calculate Field.

2) Confirm that BlockGroups is set as the Input Table parameter and Pop Density as the Field Name parameter.

3) In the list of Fields, double-click Total Pop to add it to the expression text box under POPDENSITY = . Field names appear surrounded with exclamation points (!).

Note that the field aliases you defined earlier are displayed in the list, but the field names are displayed in the expression box.

4) Under the list of Helpers, click the Division symbol (/) button.

5) In the list of fields, double-click Sq Miles. The complete expression should appear as follows:
   `!TOTPOP_CY! / !SQMILES!`

6) Compare your tool to the figure and click Run.

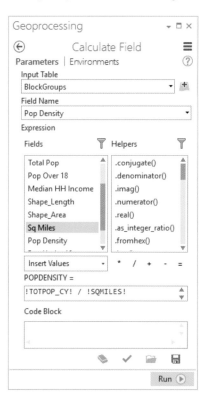

166   Lesson 4: Build the database

ArcGIS Pro calculates the population density for each record and adds the values to the table.

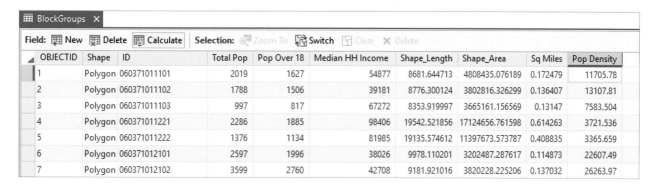

Add fields to calculate the percentage under 18

To get the percentage of the population under 18, you'll subtract the 18-and-over population from the total population and divide the result (the under-18 population) by the total population. Again, you'll need two new fields to hold your calculations.

1) Click the Add Field button at the top of the attribute table to open the fields view.

2) Add a field with following properties:
   - Name: POPUNDER18
   - Alias: Pop Under 18
   - Data Type: Short

3) In the Fields ribbon at the top of the application, click New Field.

4) Add a field with the following properties:
   - Name: PCTUNDER18
   - Alias: % Under 18
   - Data Type: Float

5) Compare your Fields pane to the figure, and click Save on the Fields tab.

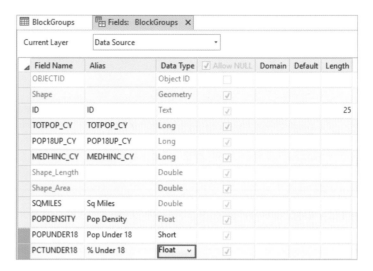

6) Close the Fields pane and return to the BlockGroups attribute table. Note that the two new fields now appear in the attribute table.

Calculate the fields

Now that you have new empty fields, you'll calculate values to fill them with data.

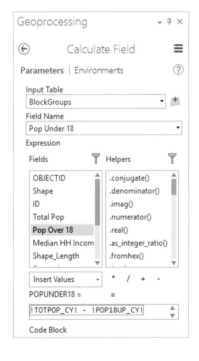

1) In the attribute table, right-click the Pop under 18 field heading and click Calculate Field to open the tool.

2) Confirm that BlockGroups is selected for Input Table and Pop Under 18 is selected for Field Name.

3) In the list of fields, double-click Total Pop to add it to the expression text box.

4) Under the list of Helpers, click the minus sign.

5) In the list of fields, double-click Pop Over 18. The finished expression should read as follows:

   `!TOTPOP_CY! - !POP18UP_CY!`

6) Compare your tool to the figure and click Run.

ArcGIS Pro calculates the population under age 18 for each record.

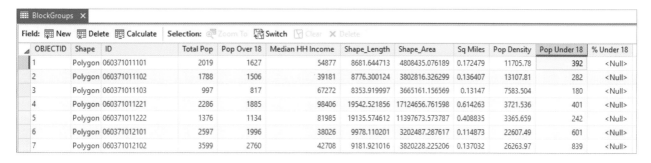

The last block group calculation you must make is to divide the under-18 population by the total population to get the percentage of children. You have a potential problem with records that have a total population of zero.

7) Scroll back across the table to see the Total Pop field.

If you scrolled down the records, eventually you would find at least one with a population of zero. Block groups can occasionally be uninhabited. Since division by zero is undefined, your calculations will give you an error message. Instead, you'll use an attribute query to select the records with nonzero values in the Total Pop field. Then you'll run the calculation on just those records.

8) On the Map tab, in the Selection group, click the Select By Attributes button.

9) In the Select Layer By Attribute tool, confirm that BlockGroups is selected as the Layer Name parameter.

10) Add a clause and fill it out to select features where the total population is greater than 0.

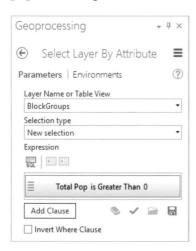

11) Compare your tool to the figure in step 10 and click Run.

Exercise 4f: Prepare the block group data

12) Look at the BlockGroups attribute table to see that 6,404 of 6,417 rows are selected (displayed at the bottom of the table).

13) Close the Geoprocessing pane.

Calculate another field

1) Scroll to the end of the table, right-click the % Under 18 field heading, and click Calculate Field.

2) Confirm that BlockGroups is selected for Input Table and % Under 18 is selected for Field Name.

3) In the expression text box, create the following expression: `(!POPUNDER18! / !TOTPOP_CY!) * 100`.

▶ Make sure you include the parentheses as shown.

4) Compare your expression to the figure and click Run.

Values are calculated for the selected records. The unselected records still have null values. You'll calculate these values to zero, because if no one lives in the block group, it's reasonable to say that the percentage of children is zero.

5) At the top of the attribute table window, click the Switch Selection button . You should have 13 of 6,417 records selected.

6) Right-click the % Under 18 field heading again and click Calculate Field.

7) In the expression text box (under PCTUNDER18 =), clear any text and type 0.

8) Click Run.

9) At the top of the table, click the Clear Selection button

You now have the attributes you need for population density and percentage of the population under age 18.

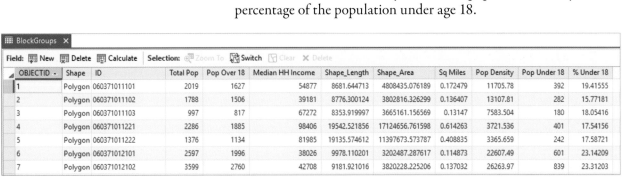

Calculate Summary Statistics

Because these calculated values will be used in the analysis later in the project, you must ensure the correct values have been calculated in the BlockGroups table.

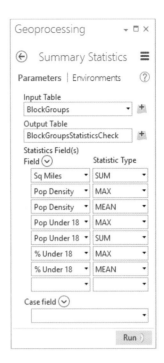

1) Use the Summary Statistics tool, accessed through the Tools button on the Analysis tab, to calculate the following statistics for BlockGroups:

    - Sq Miles: Sum
    - Pop Density: Max and Mean
    - Pop Under 18: Max and Sum
    - % Under 18: Max and Mean

2) Name the Output Table BlockGroupsStatisticsCheck.

3) Compare your table to the figure, and confirm that all your values match. Then click Run.

4) Open the table (under Standalone Tables in the Contents pane) for BlockGroupsStatisticsCheck and view the new fields.

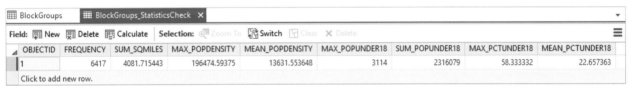

5) Confirm your calculated values. Your values may be slightly different out to the last decimal places.

    - Sq Miles   4,081.715443
    - Max Pop Density   196,474.59375
    - Mean Pop Density   13,631.553648
    - Max Pop Under 18   3,114
    - Sum Pop Under 18   2,316,079
    - Max % Under 18   58.333332
    - Min % Under 18   22.657363

6) Close the tables.

7) Save your project.

Exercise 4g: Prepare the parcel data

Your search for suitable park sites is limited to vacant land parcels. As you saw in lesson 2, vacancy is an attribute of the stand-alone, nonspatial Vacant Parcels table. You must associate this attribute with the Parcels feature class so that you can see and select vacant parcels on the map. You'll do this with an operation called a *table join*. A table join attaches the attributes of one table (usually, a nonspatial stand-alone table) to those of another (usually, a feature class attribute table). The join is based on a field of values, typically an identification code, which is common to both tables. Values in the common field are used to match records in the stand-alone table with records in the feature attribute table. For more information, see the sidebar "Table joins and relates" later in this lesson.

Add data

First, you'll add Parcels and Vacant Parcels to the map.

1) If necessary, start ArcGIS Pro and open the LARiver_ParkSite project.

2) Make a copy of the Lesson4f map and rename it Lesson4g.

3) Open the Lesson4g map.

4) Under Contents, turn off all the layers except LARiver, Counties, and the Topographic basemap to simplify the map display.

5) Expand Folders in the Project pane and browse to SourceData > City of LA. Then add Parcels.shp and VacantParcels.dbf to the map.

Because VacantParcels.dbf is simply a table and not a layer with vector features stored with it, the table is added to the Standalone Tables group in the Contents pane.

6) Zoom to the Parcels layer.

Identify the common attribute

You'll look at both tables to find the common attribute, or key field, to form the join.

1) Open the Parcels attribute table and then the Vacant Parcels table.

2) At the top of the table window, drag the Vacant Parcels tab to the blue docking target on the right.

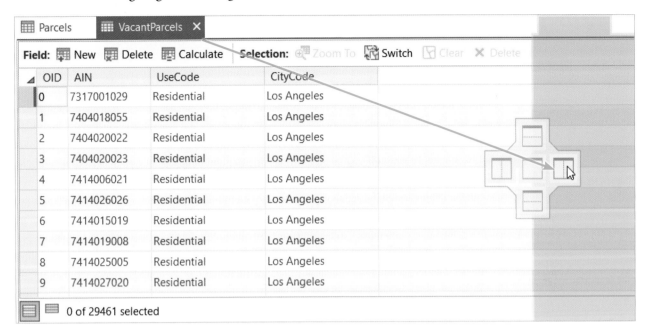

3) In the Parcels table and the Vacant Parcels table (on the right), note that both display an AIN (assessor identification number) attribute.

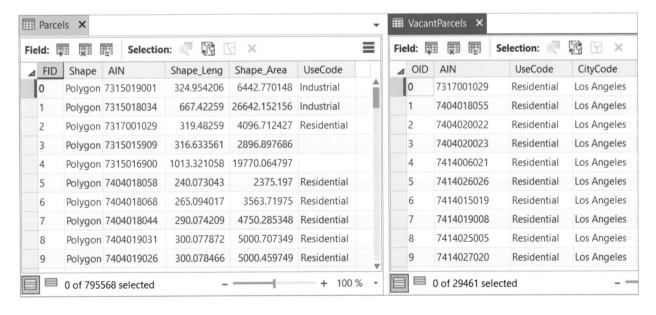

The AIN is a unique identifier. In theory, when you join the tables, each of the 29,461 records in the Vacant Parcels table should find a single matching record in the Parcels table.

For more information, see the sidebar "Table joins and relates."

Join tables

You don't need to have the tables open to join them.

1) **Close the two table windows.**

2) **Under Contents, right-click the Parcels layer and click Joins and Relates > Add Join.**

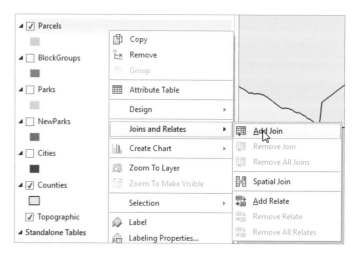

The Geoprocessing pane opens with the Add Join tool displayed.

3) **Confirm that Parcels is set as the Layer Name parameter and Vacant Parcels as the Join Table parameter.**

4) **In the Input Join Field drop-down list, click AIN. This specifies the key field in the Parcels table.**

When you select the Input Join field, ArcGIS Pro will automatically look for the field with a matching name in the join table. Because you have an AIN field in both the input and join layers, AIN is matched for you as the Output Join field.

5) **Confirm that AIN is selected for Output Join Field.**

6) **Confirm that drop-down list number 2, for Join Table, is set to VacantParcels.**

Table joins and relates

Tables can be associated in ArcGIS Pro through either a join or a relate. A join attaches one table's attributes to the other table. (The join is "virtual," meaning that it exists only in the map document; the two tables remain separate on disk.) A relate is a look-up relationship, in which you select records in one table to see matching records in the other table. A join is stronger than a relate because attributes in a joined table can be used to symbolize and query features on the map, whereas attributes from a related table cannot. Both joins and relates require a common attribute to match records.

A join is appropriate when, for any given record in the feature class table (the table to which attributes are being attached), there is at most one possible match in the stand-alone table (the table from which attributes are being attached). A relate is appropriate when more than one match is possible. Consider some examples:

1) A feature class of parcels and a stand-alone table of vacant parcels, in which the common attribute is Parcel ID, a unique identifier in both tables. Each record in the Parcels table has, at most, one matching record in the Vacant Parcels table so a join may be used. The records have a one-to-one relationship: any given record in either table can have no more than one match in the other table. (Not all records in the Parcels table will have matches because not all parcels are vacant.).

2) A feature class of rivers and a stand-alone table of watersheds, in which the common attribute is Watershed ID. This is a nonunique identifier for rivers, because all rivers that drain into the same watershed, or drainage area, have the same Watershed ID. However, it is a unique identifier for watersheds. Because a river drains into just one watershed, each record in the Rivers table has, at most, one matching record in the Watersheds table, which means that a join may be used. The records have a many-to-one relationship: many records in the feature class table can match the same record in the stand-alone table.

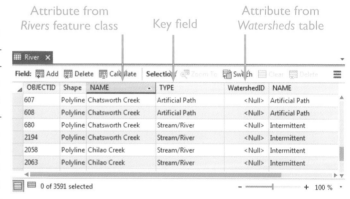

3) A feature class of the Los Angeles River and a stand-alone table of improvement projects, in which the common attribute is River ID. This is a unique identifier for river features. It is a nonunique identifier for the stand-alone table, because more than one improvement project may be located along the same river. In this situation, a relate should be used because a given record in the feature class table may have more than one match in the stand-alone table. In a join, only one matching record could be attached to the feature class table; the others would be left out. The records have a one-to- many relationship: one record in the feature class table may have many matches in the stand-alone table.

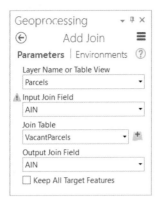

7) **Click to clear the Keep All Target Features check box.**

The only parcels that interest you are the vacant ones—those with matches in the VacantParcels table. If this box is checked, the layer will have all the parcels, with null values for those records that can't be matched.

8) **Compare your tool to the figure and click Run.**

When the tool is finished running, the Parcels layer redraws on the map. Because you chose to keep only the matching records, vacant parcels are the only ones that draw.

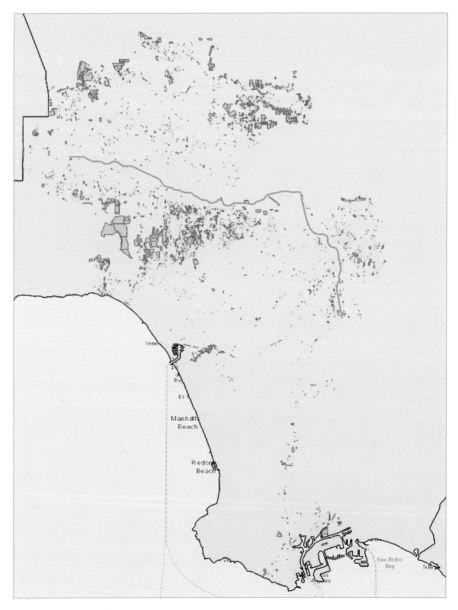

9) Open the attribute table of the Parcels layer.

10) Examine the attribute fields.

The joined table has the attributes of both tables. It has 29,461 joined records, which means that all the records from the Vacant Parcels table were matched to the Parcels layer.

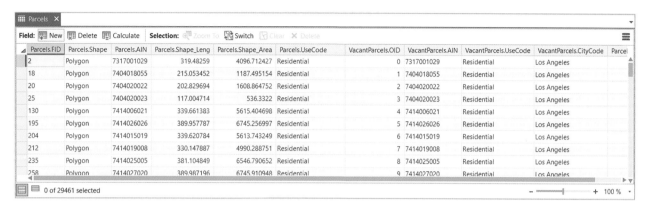

11) Close the table.

Copy features

It's important to understand that a join does not actually save the new fields to the input feature class. The join is maintained "virtually" for only this layer in this particular map. You'll now create a new feature class in the project geodatabase, copying just the vacant parcel features from the joined table.

1) Open the Copy Features tool.

2) In the Copy Features dialog box, set Input Features to the Parcels layer.

3) In the Output Feature Class box, type VacantParcelsJoin.

4) Compare your tool to the figure and click Run.

When the tool is finished running, the new feature class appears in your project geodatabase, and a layer is added to the map.

5) Under Contents, remove the Parcels layer and the Vacant Parcels table.

6) Open the attribute table of the VacantParcelsJoin layer and examine the attribute fields.

The virtual attributes of the joined table are now permanent attributes of the new feature class.

Exercise 4g: Prepare the parcel data   177

Field names from the two original tables in the join have been converted to aliases in this table.

7) **Using the Menu button, you can turn off the aliases to see the complete field name that indicates with an underscore which source table the attribute came from.**

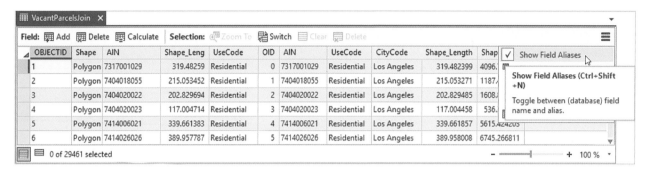

Dissolve vacant parcels

You must still calculate acreage for the vacant parcels (row 3 of the data requirements table). At this point, you might have a new thought: What if there are some adjacent vacant parcels that are smaller than a quarter acre individually but would be more than a quarter acre if combined? You might be missing out on some potential park sites simply because of property lines.

This figure shows two adjacent vacant parcels. Separately, each is less than a quarter acre, but together they're more than a quarter acre. You'll want to combine them into a potential park site.

To address this possibility, you'll dissolve the VacantParcelsJoin feature class, which will create single features out of parcels sharing a common boundary. In the process, you're going to lose attributes. You don't really need the attributes though—at least, you don't need them as much as you need good park locations.

1) On the Analysis tab, open the Dissolve tool.
2) In the Dissolve dialog box, set Input Features to the VacantParcelsJoin layer.
3) Change the Output Feature Class name to VacantParcels.
4) Leave the Dissolve_Field(s) area blank.

You're dissolving purely on the parcel geometry. See the sidebar "Dissolving features" earlier in this lesson for a discussion on dissolving on geometry versus dissolving by attributes.

5) At the bottom of the dialog box, click to clear the Create multipart features check box.

If you allowed multipart features, the output feature class would consist of a single multipart polygon. You'd have one great big discontinuous vacant parcel.

6) Compare your tool to the figure and click Run.

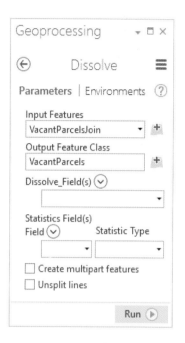

When the dissolve is completed, VacantParcels is added to the project geodatabase, and a layer is added to the map.

7) Under Contents, turn VacantParcels off and on a few times.

Even with the map zoomed out to the city limits, you can see some areas where parcel boundaries have disappeared.

8) Under Contents, remove the VacantParcelsJoin layer.
9) Open the attribute table of the VacantParcels layer.

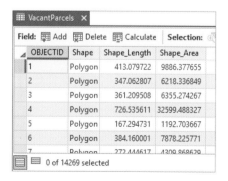

It has 14,269 records, about half as many as the undissolved layer. The only attributes present are the four managed by ArcGIS Pro.

Calculate acreage

Now you can get back to acreage. ArcGIS Pro has recalculated the Shape_Area field for the dissolved parcels, but the units are in the units of the map projection. You need to convert them to acres.

1) Add a new field to the VacantParcels table.

2) Give the new field these properties:
   - Name: ACRES
   - Alias: Acres
   - Data Type: Double

3) On the Fields tab, click Save.

4) Close the fields view.

5) Back in the table, right-click the Acres field heading and click Calculate Field.

6) Type the following text in the text box under ACRES =:

   `!shape.area@acres!`

7) Compare your tool to the figure and click Run.

ArcGIS Pro calculates the acreage for each record and adds the values to the table.

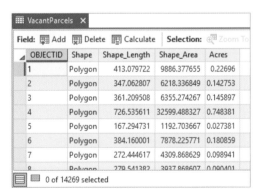

180   Lesson 4: Build the database

8) Use the Summary Statistics tool to calculate the total sum of vacant parcels in acres and the average size. Name the Output Table parameter VacantParcelsAcresSumMean.

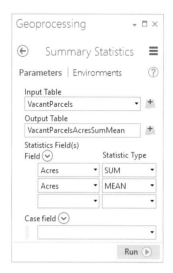

You should have close to 17,244.6 total acres and an average parcel area of 1.2 acres.

9) Close all the tables.

Clean up the project geodatabase

You'll get rid of the undissolved VacantParcelsJoin, and then you'll edit the item description for this VacantParcels feature class.

1) In the Project pane, in the project geodatabase, right-click VacantParcelsJoin and click Delete.

▶ Make sure VacantParcelsJoin—not VacantParcels—is highlighted in the Project pane.

2) On the prompt, click Yes to confirm your deletion. The feature class is deleted from the geodatabase.

3) In the Project pane, right-click the VacantParcels feature class and click View Metadata.

Except for the title (which reflects the feature class name) and the file format (type) at the top, the item description is empty.

4) On the Home ribbon, in the Metadata group, click Edit.

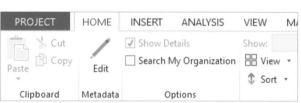

5) In the Title box, change the title to Vacant Parcels (with a space).

6) Scroll down and click in the Tags box. Type the following: parcels, vacant parcels, Los Angeles.

▶ Include the commas: they separate tags.

7) Scroll down, if necessary. Click in the Summary box and type: This dataset of land parcels in the city of Los Angeles supports queries on the size of contiguous vacant parcels.

8) Click in the Description box and type: This dataset represents 29,461 original vacant parcels in the city of Los Angeles. Adjacent parcels have been dissolved, leaving 14,269 features.

9) Click in the Credits box and type: City of Los Angeles.

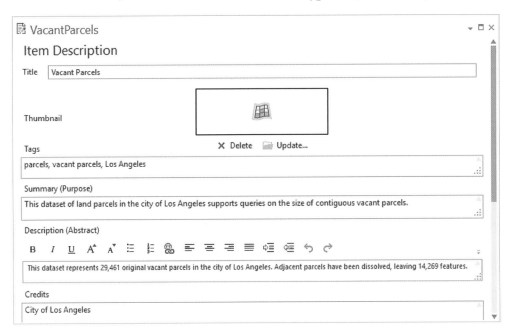

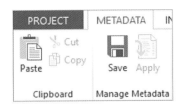

10) On the Metadata ribbon, in the Manage Metadata group, click the Save button.

11) Close the Project tab, all the Lesson4 tabs, and any open tables.

12) In the Project pane, confirm that your project geodatabase has the feature classes and tables shown at right.

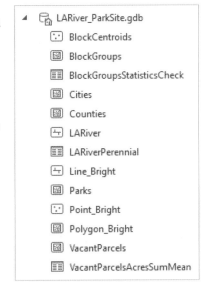

13) Save your project.

14) Close the data requirements table, if you have it open.

15) Continue to the next lesson or close ArcGIS Pro. Save your changes if prompted.

Lesson

5 Edit the data

DATA INTEGRITY IS CRUCIAL IN GIS,

and ArcGIS Pro has some built-in safeguards to make sure you really mean to be doing what you're doing. When editing, if you make a mistake, you can recover with the Undo button, which lets you undo edits one by one in reverse order. If the mistakes are too complicated to resolve, you can discard your edits without saving them.

Ideally, data would come to us free of errors or inconsistencies, but that seldom happens. Careful exploration of data usually reveals imperfections, which arise for all kinds of reasons. Sometimes data is captured or created incorrectly to begin with. Sometimes errors creep in with subsequent data processing. Sometimes the data is perfectly good until the world changes. And sometimes the data isn't wrong at all—it's just generalized for use at a certain scale and inappropriate for use at other scales. For example, the park data you're using was really designed for medium-scale maps, more or less in the range of 1:100,000 to 1:250,000. When the features are examined at much larger scales, it's not surprising that they don't conform to high-resolution imagery.

ArcGIS Pro has many tools for editing features in specialized ways, as well as methods for evaluating and maintaining data integrity. Data editing, however, isn't the focus of your project. You're correcting one park feature and creating another because, during your data exploration, you saw a need to do so. Then you'll move ahead with the analysis.

Exercise 5a: Edit a feature

In lesson 3, you saw that Pecan Playground, a park near Dodger Stadium, didn't match up well against the imagery. In this exercise, you'll edit the park boundary and update its acreage attribute.

Lesson Five road map

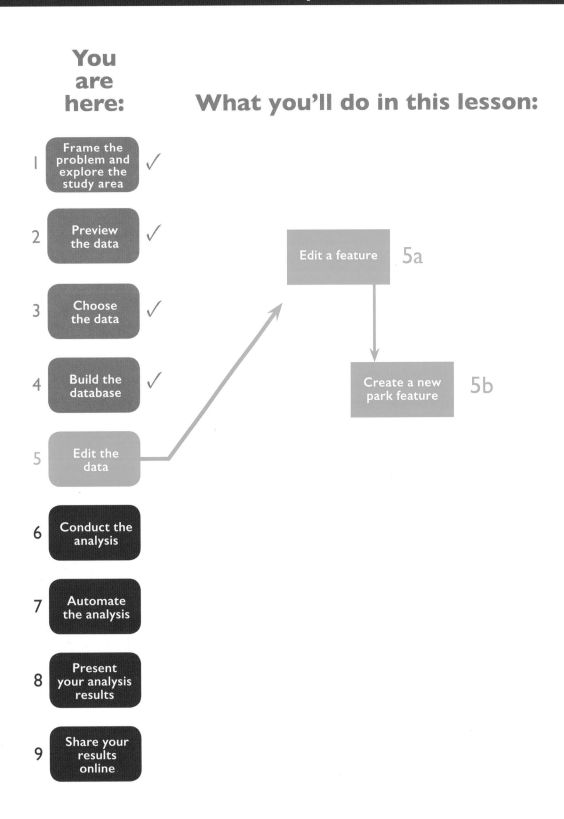

Add data to ArcGIS Pro

You need the park data and a basemap of imagery.

1) Start ArcGIS Pro and open your LARiver_ParkSite project.
2) Insert a new map and name it Lesson5.
3) Change the basemap to Imagery.
4) In the Project pane, navigate to the LA River project geodatabase in the Databases folder.
5) From the LARiver_ParkSite geodatabase, drag Parks to the map. If necessary, right-click the Parks layer and click Zoom To Layer.

Zoom to Pecan Playground and edit

1) Zoom to the Pecan Playground bookmark.

You should change the park symbol so you can see what you're doing as you're editing.

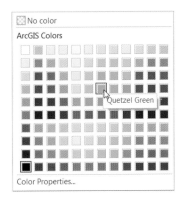

2) Open the Symbology pane for the Parks layer and change the fill color to No Color.

3) Change the outline color to Quetzel Green and the outline width to 2. Click Apply.

4) Zoom and pan as needed so you see the whole block, as in the figure on the lower left.

The feature as drawn doesn't include the swimming pool or the play area at the north end of the block. Also, its boundaries extend into the street. (The southwest corner of the block is correctly excluded as this is a school.) After editing, Pecan Playground should look more like the modified version.

Swimming pool   Playground

Basemap layers are updated periodically, so your imagery may look different.

Don't worry too much about your results—it should be fairly easy to make the feature better than it is now, and that's good enough.

5) On the Edit tab, click the Select button. Click anywhere inside the Pecan Playground feature to select it. The feature outline turns bright blue momentarily.

6) Confirm that snapping is turned on. The Snapping tool on the Edit tab should be blue. If not, just select it to turn it on.

7) **Click the Modify button in the Features group. The Modify Features pane opens.**

8) **Under Reshape, click the Vertices tool.**

Two things happen. One is that a new toolbar, the Edit Vertices toolbar, is added.

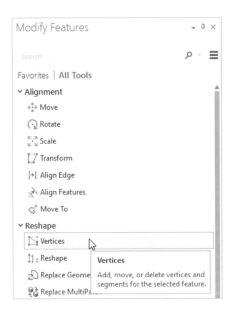

The other is that the feature is now marked with a number of small green squares (and one red one). The squares, or vertices, are coordinate pairs that define the feature's shape and position. The red vertex is the last one added when the feature was created or edited. In the figure, though not on your screen, the vertices are numbered for reference. Note that the red circles with numbers will not appear on your map. They are for visual instructions only.

As you edit the feature, you may find it useful to zoom and pan using shortcut keys.

Shortcut keys for navigating while editing

To navigate while editing vertices, use these shortcut keys: Z to zoom in; X to zoom out; C to pan; B for zoom/pan (click and drag to zoom; right-click and drag to pan). Clicking tools such as Explore interrupts vertex editing. If you accidentally interrupt vertex editing, click the Vertices button on the Modify Features pane to resume it.

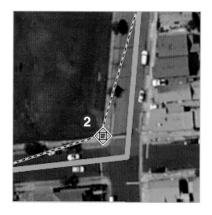

9) On the Edit Vertices toolbar, make sure the Normal tool is selected.

10) Place your pointer over vertex 2, which is in a street intersection.

When the pointer is directly over the vertex, it changes to a four-headed arrow.

11) Click and drag the vertex to where it should go—the southeastern corner of the soccer field—and then release the mouse button.

▶ If you make a mistake, you can recover with the Undo button ↶ on the Quick Access toolbar.

The blue highlight continues to show the original shape of the feature.

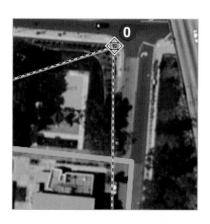

12) Place your pointer over vertex 0 and move the vertex to the northeastern corner of the park.

The eastern boundary now has a kink because of vertex 1. This vertex is superfluous.

13) On the Edit Vertices toolbar, click the Delete tool.

14) Click on vertex 1 to delete the vertex.

The eastern side of the park should now be straight, as in the figure.

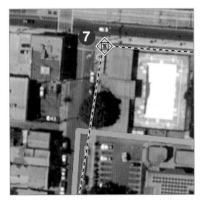

15) On the Edit Vertices toolbar, click the Normal tool and move vertex 7 (the red vertex) to the northwest corner of the park.

16) Move vertex 5 to where the baseball diamond meets the schoolyard.

17) Click the Delete tool and delete vertex 6 to straighten the western side of the park.

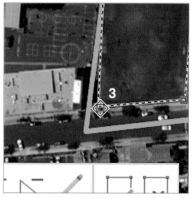

18) Click the Normal tool and move vertex 3 to the southwestern corner of the soccer field.

Explore snapping

When you move a vertex near another feature, the vertex connects to it automatically. This behavior, called *snapping*, helps prevent errors such as small gaps and overshoots. A vertex will snap to features in its own layer as well as to features in other layers. You may have seen this behavior already with one or two of the vertices you've moved.

1) Hover over vertex 4, at the northeastern corner of the school.

This vertex is already at a good location. You're just going to explore snapping behavior a little bit.

2) **Without releasing the mouse button, drag the vertex a small distance in various directions.**

As you move south or west along the edges of the feature, you'll see the SnapTip "Parks: Edge" appear next to the pointer. The SnapTip tells you that you're within the range, or *snapping tolerance*, of an edge in the Parks layer. As you move back toward your original position, you'll see the SnapTip "Parks: Vertex." Moving in any direction away from the feature, you won't see a SnapTip.

3) **Drag the vertex back toward its original position. When you see the "Parks: Vertex" SnapTip, release the mouse button.**

4) Use your scroll wheel to zoom in on vertex 4.

▶ If you don't have a scroll wheel, press and hold the Z key, and then drag the mouse to zoom in and out. If you zoom in too far and the imagery disappears, press and hold the X key and zoom out a bit.

5) **Place your pointer over vertex 4 again. Without releasing the mouse button, drag the vertex a small distance in different directions.**

The snapping behavior is the same as before, but the snapping tolerance—which is 10 screen pixels by default—now corresponds to a much smaller ground distance. That gives you more control over the exact placement of the vertex.

6) **Again, drag the vertex back toward its original position.**

When you see the "Parks: Vertex" SnapTip, release the mouse button.

7) **Zoom out to see the entire diagram.**

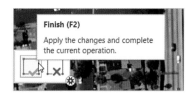

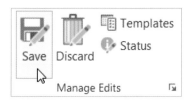

Save your edits

You'll finish the diagram and save your edits.

1) **On the Edit Vertices toolbar, click the Finish button.**

The edited park is now a selected feature. You could still undo these edits, or stop the edit session without saving them, but you should be satisfied with the new park boundary.

2) **On the Edit tab, in the Manage Edits group, click Save.**

▶ Saving your project does not save your data edits. You must click the Save button on the Edit tab to save the changes to the database.

Update attribute values

Changing the park's shape has changed its area. That means the ACRES attribute value is no longer correct.

1) **Open the Parks attribute table.**

2) **At the bottom of the table, click the Show Selected Records button. Confirm that only Pecan Playground is selected. If necessary, scroll across to see the geometry attributes.**

ArcGIS Pro has automatically updated the Shape_Length and Shape_Area fields in the linear unit of the data's projection, but it's your job to maintain the ACRES attribute. The value stored now is just over four acres (4.1). This is the acreage of the old, pre-edited feature.

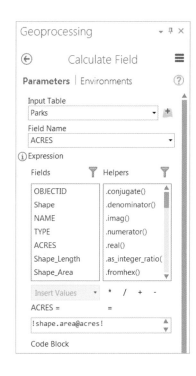

3) **Right-click the ACRES field heading and click Calculate Field.**

4) **Enter the following text in the text box under ACRES =:**

   `!shape.area@acres!`

5) **Compare your tool to the figure and click Run.**

The new value is written to the table. It shouldn't be much different from the old value. Although the park was enlarged along its northern boundary, it was narrowed along three sides.

6) **At the bottom of the table, click the Show all records button.**

7) **On the Edit tab, click the Clear Selection button.**

8) **Close the Parks attribute table.**

9) **Close the Modify Features pane.**

Your edits have been saved to the Parks feature class and are preserved whether you save the project or not. You'll save the project because you'll keep using it in exercise 5b, and you may not be continuing immediately.

If you are continuing to the next exercise, leave ArcGIS Pro open; otherwise, close ArcGIS Pro.

Exercise 5b: Create a new park feature

Another local park, Vista Hermosa Park, was not represented by a feature in any of the parks feature classes. In this exercise, you'll create it by tracing its outline against the basemap image. The process is similar to the editing you did in exercise 5a.

After creating the feature, you'll add attribute values for it. You'll also edit the attribute values of the two park features you loaded in lesson 4.

Start editing

You'll zoom to the location of the missing park and start editing.

1) **If necessary, start ArcGIS Pro and open the Lesson5 map.**
2) **Use your bookmarks to zoom to Vista Hermosa Park.**

The park itself doesn't take up the entire L-shaped lot, just the portion shown by the dotted outline in the figure.

You won't see the dotted line on your map—it was added here for reference.

3) **Zoom in to an area more or less matching the area around the park.**

4) **On the Edit Tab, in the Features group, click Create.**

5) **In the Create Features pane, click the Parks symbol.**

By clicking the symbol, you've chosen a template. A template determines the properties that new features you create will have. Because your map has only one editable layer, Parks is the only template available. And because all your parks are symbolized the same way, the template has just one symbol.

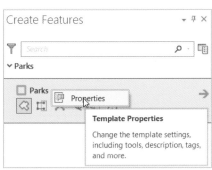

6) **Right-click the Parks layer in the Create Features pane and click Properties.**

The Template Properties dialog box lets you view and change the properties of the template. You can switch between General, Tools, and Attributes.

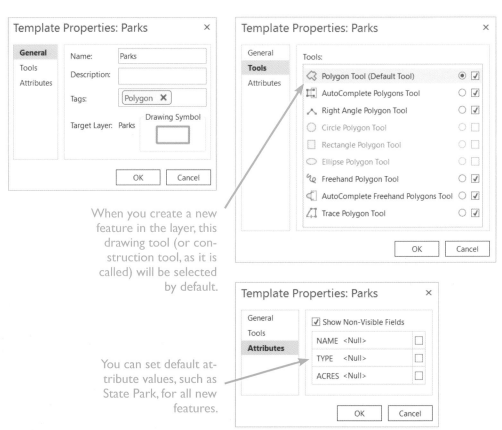

When you create a new feature in the layer, this drawing tool (or construction tool, as it is called) will be selected by default.

You can set default attribute values, such as State Park, for all new features.

You don't need to change any of these settings.

7) **Close the Template Properties dialog box.**

Feature templates are especially useful when you're creating many new features in a layer with varied symbology. For example, a Points of Interest layer might have different symbols and descriptive attributes for restaurants, museums, parks, and so on. A template allows you to choose the right symbol and description for each new point as you add it.

Create the park feature

You're ready to draw the boundary of Vista Hermosa Park. Again, your purposes don't require extremely high accuracy. You just want a boundary that approximately conforms to the yellow dotted outline in the figure at the beginning of this exercise. If you decide to zoom in and work at large scale, you may be able to see and follow the fence line that encloses the park. (Remember to use the editing shortcut keys.)

1) **Move your pointer over the map. The cursor becomes a cross hair.**

2) **Choose a starting point.**

Your starting point can be anywhere, although corners are usually good—for example, you might use the southeast corner of the soccer field.

3) **Click once to start drawing the feature. A red vertex is added at the location where you clicked.**

4) **Move the mouse in the direction you want to go. (Don't drag, just move.)**

As you move, the cursor remains connected to the vertex by a purple line. As long as you're moving in a straight line, you don't need to add a vertex.

5) **When you come to a spot where you must change direction, click to add a second vertex.**

The new vertex is red (it's the last one added), and the first vertex turns green.

6) **Trace the boundary of the park as best you can, following fence lines, property lines, and sidewalks.**

   ▶ If you add a bad vertex, click the Undo button on the Quick Access toolbar.

7) **When the polygon is complete, click the Finish button on the Drawing toolbar (or simply double-click).**

Your new feature, still selected, should resemble the figure.

Again, the result doesn't have to be perfect. However, if you want to make changes, click the Vertices button in the Tools gallery. If you're really dissatisfied with your feature, press the Delete key on the keyboard to get rid of it, and then start over.

8) **On the Edit tab, click Save to save your edits.**

9) **Close the Create Features pane.**

Edit attribute values

You've created the Vista Hermosa Park feature, but you haven't given it attribute values.

1) On the Edit tab, in the Selection group, click the Attributes button ▦. Confirm that the Vista Hermosa Park feature is still selected.

The Attributes pane opens, showing the attributes and values for this feature.

Values in software-managed fields (Shape_Length and Shape_Area) are dimmed and can't be edited. (Your length and area values won't match those in the figure. Your OBJECTID value may be different, too.)

2) In the Attributes pane, click the null value next to NAME and enter Vista Hermosa.

3) Click the null value next to TYPE and enter Local park or recreation area.

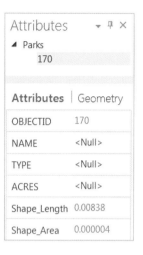

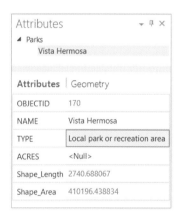

4) Close the Attributes pane.

5) On the Edit tab, click Save to save your edits.

You must still calculate the acreage for Vista Hermosa Park and for the two park features you loaded in lesson 4 (which didn't have an acreage attribute). You could find these records by scrolling to the bottom of the Parks attribute table, but a more systematic way is to query for null values in the ACRES field.

6) Open the Parks attribute table.

7) On the Map tab, in the Selection group, click Select By Attributes.

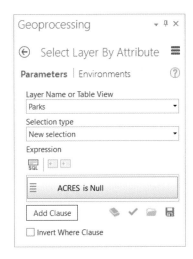

In the Select Layer By Attribute dialog box, you'll build a query expression to find parks for which the acreage is not given.

8) Make sure Selection type is set to create a new selection.

9) Under Expression, click Add Clause.

10) In the Field drop-down list, click ACRES.

11) In the operator drop-down list, click is Null.

12) Click Add.

13) Confirm that Invert Where Clause is unchecked.

14) Confirm that your expression matches the figure on the left and click Run.

15) Close the Select Layer By Attribute pane.

16) In the table, click the Show selected records button. Three records should be selected.

17) In the table, calculate the values for the ACRES field.

▶ How? Right-click the ACRES field heading and click Calculate Field. Enter the following text in the textbox under ACRES =:

`!shape.area@acres!`

Then click Run.

The results should be about 32 acres for Los Angeles State Historic Park, 36 acres for Rio de Los Angeles State Recreation Area, and 9 to 10 acres for Vista Hermosa Park.

The TYPE values for Los Angeles and Rio de Los Angeles (the two parks you loaded) are nonstandard in your table because they came from another data source. All other state parks have the value "State or local park or forest."

18) **In the table, right-click on the gray square next to the Vista Hermosa record and click Select/Unselect.**

Two records are left selected.

19) **Right-click the TYPE field heading and click Calculate Field.**

20) **In the expression box under TYPE =, type "State or local park or forest".**

▶ The quotation marks are required for a string value.

21) **Compare your dialog box to the figure and click Run.**

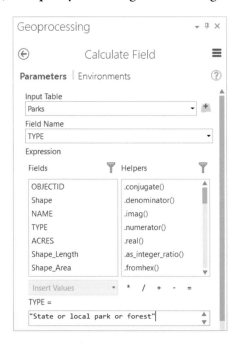

The values for the two selected records are updated.

Exercise 5b: Create a new park feature   199

22) Confirm in the attribute table that the values for the two selected records are updated.

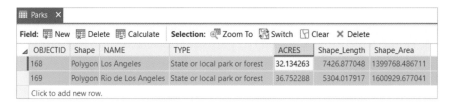

23) In the table, clear the selection and show all records.

24) Close the attribute table and Calculate Field pane.

Zoom to Parks and save the map

You'll take a quick look at your work and save your project.

1) Resymbolize the Parks layer to give it a solid green fill color and a 1 pt. outline width. Change the outline color, too, if you want.

2) Use bookmarks to zoom to Pecan Playground. Your updated boundary should be displayed.

3) Zoom to the Vista Hermosa Park bookmark. Your park should be displayed.

As mentioned earlier, edits are written to the source data at the time you save them. Although you're making a practice of saving maps, your park edits have already been saved to the feature class.

4) Close the Lesson5 tab and any open tables.

5) Save your project.

6) Continue to the next lesson or close ArcGIS Pro. Save your changes if prompted.

All your data preparations are complete. In lesson 6, you'll look for suitable park sites.

Lesson

6 Conduct the analysis

NOW WE COME TO THE POINT OF your preparatory work: the analysis. Your original feature class of parcels was like an assembled 796,000-piece jigsaw puzzle covering the entire city, with each piece representing an individually owned property. In lesson 4, you cut this down to about 29,000 parcels (those that were vacant) and then to around 14,000 (those that were adjacent). Now you'll continue taking pieces away, removing unsuitable parcels until you're left with only those candidates that meet the criteria established in lesson 2:

On a vacant land parcel at least one-quarter acre in size that is

- within the LA city limits,
- within half a mile of the LA River (the closer the better),
- at least a quarter mile from the nearest park,
- in a neighborhood in which
    - population density is at least 8,500 people per square mile,
    - at least 22 percent of the population is under 18, and
    - median household income is $50,000 or less, and
- preferably serving the most people within a quarter-mile radius.

But how does this translate into a method and workflow? How do you take these requirements and match them to appropriate geoprocessing tools in the right sequence? Planning a workflow means identifying the tasks, finding a suitable tool for each one, and understanding the inputs and outputs so you can order the operations in a logical, efficient way.

Most of the tasks in your analysis—and this is typical—fall into just a few categories: distance problems (what's near what), topological problems (what crosses, touches, or contains what), attribute query problems (what has this or that value), and overlay problems (what areas are common to features in different layers). Each of these problem categories is associated with its own special set of tools.

Before beginning an involved analysis, you may find it helpful to draw a rough plan or diagram. Any medium will do—paper, an app, a whiteboard—as long as you're prepared to make revisions as you

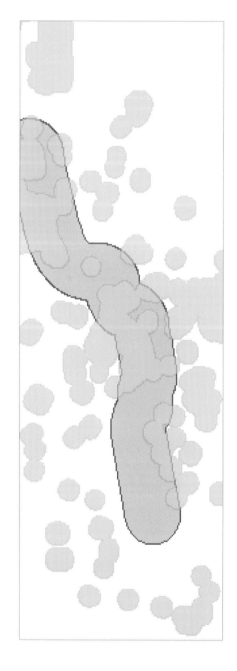

Lesson Six road map

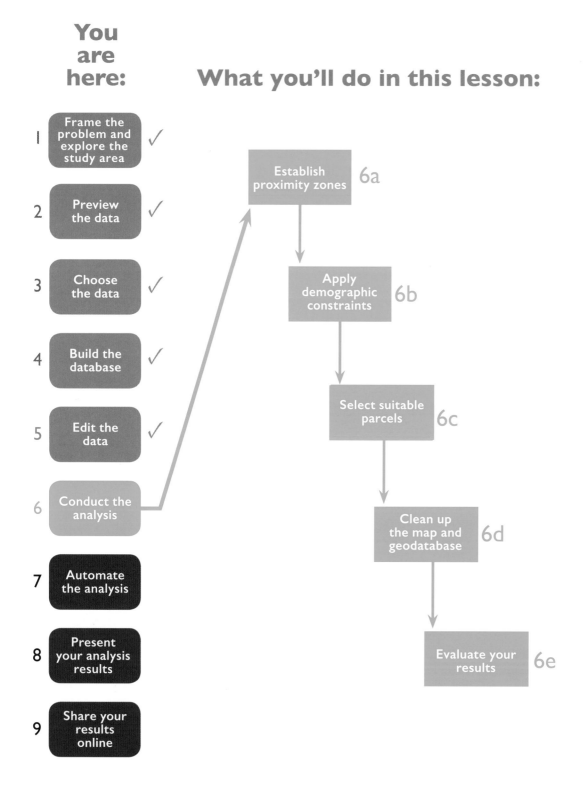

go. A typical working plan (the one you'll follow) could be sketched out to look something like the diagram here. This sketch depicts the broad general approach, although there are bound to be a few additional twists and turns along the way.

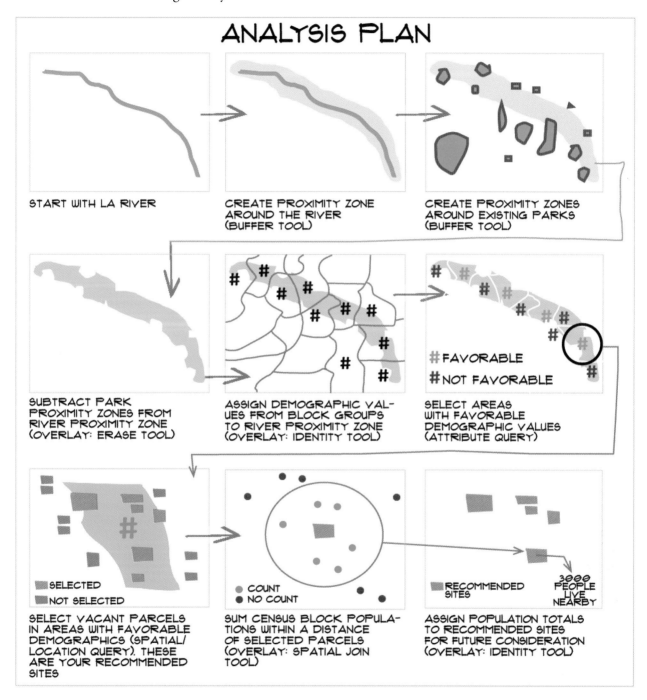

Looking at the sketch of the analysis, you can see that this plan represents only one of several possible approaches to the problem. The plan you're following here is, we think, a reasonable and efficient one. It uses a variety of important geoprocessing tools and has the visual benefit of letting you see the analysis unfold, as land areas are progressively stripped away from consideration. It is certainly possible, however, to reach the same or similar conclusions using different combinations of tools.

Your analysis results depend on the state of the data in the Project geodatabase. If you've successfully completed all the exercises up to now, you should be in good shape. If necessary, download the lesson 5 results from the book resource web page, at esri.com/Understanding-GIS-3.

Exercise 6a: Establish proximity zones

In this exercise, you have two objectives. First, you want to define the area of interest as a half-mile zone around the river. Any parcels falling outside this zone will be excluded from consideration. Second, you want to draw quarter-mile zones around each park. Any parcels in the area of interest that lie within these internal exclusion zones will also be dropped. You'll use one simple operation—buffer—to significantly reduce your hunting grounds.

Create a results geodatabase

When you generate outputs, you need to store them somewhere. In lesson 4, you decided to keep your input data in one geodatabase and your output data in another geodatabase. (See the sidebar "Project database considerations" in lesson 4.) It's a matter of preference, but we think it's easier to keep the project organized if inputs and outputs are separated. You may want to share the input data with someone else so they can run the analysis independently. You may want to repeat the analysis with different parameters (something that you'll do in lesson 7). The more feature classes you add to a geodatabase, the harder it is to keep track of what they represent and what purpose they serve.

1) Start ArcGIS Pro and open your LARiver_ParkSite project.

2) Insert a new map to your project and name it Lesson6a.

3) Close any other open maps by closing their associated tabs above the map. You can always get back to these maps by double-clicking them in the Project pane, in the Maps group.

4) In the Project pane, expand the Databases folder. It should contain your LARiver_ParkSite project geodatabase.

5) Right-click the Databases folder and click New File Geodatabase. By default, the dialog box should already be set to the project folder so that the new geodatabase will be created in this folder.

6) Name the geodatabase AnalysisOutputs and click Save.

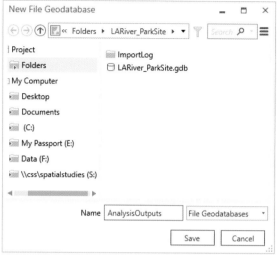

Note that the new geodatabase has been added to the list of databases in the Project pane.

You'll want to make AnalysisOutputs the current workspace so that your output data is put here by default.

Set the current workspace

When working with geoprocessing tools in ArcGIS Pro, it is convenient to set a default location to save the outputs (in this case, the new AnalysisOutputs geodatabase). This default output location is known as the "current workspace" and can be set as an environment setting for the project. The "scratch workspace" is a second environment setting option for the output location of temporary data – data that you don't need to keep and maintain. By setting both the current workspace and scratch workspace, you save yourself the step of browsing to a new output location every time you run a tool.

1) On the Analysis tab, in the Geoprocessing group, click the Environments button 🛠.

2) Click the browse button next to the Current Workspace box and browse to the new AnalysisOutputs geodatabase. Click OK.

3) Set Scratch Workspace to the same geodatabase.

4) Compare to the figure and click OK on the Environments dialog box.

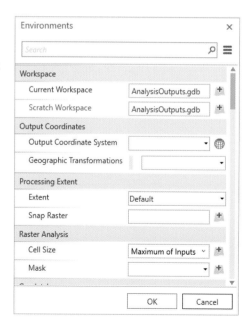

Buffer the LA River

A buffer is a zone around a map feature measured in distance units. You'll use the Buffer tool to create a half-mile proximity zone around the LA River. For more information about the Buffer tool and other analysis tools, see the sidebar "Essential GIS analysis tools."

1) **In the Project pane window, expand LARiver_ParkSite.gdb and add LARiver to the map.**

Note that the layer is likely symbolized in a color other than blue. During analysis, you're more interested in geoprocessing results than map appearance. For that reason, you're not going to symbolize input and output layers carefully at each step along the way. You'll do it as the need arises, but often you'll just accept the ArcGIS Pro defaults.

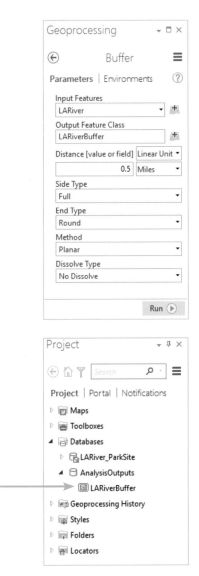

2) **On the Analysis tab, in the Tools gallery, click Buffer. The Geoprocessing pane opens with the Buffer tool parameters displayed.**

3) **In the Geoprocessing pane, click the Input Features drop-down arrow and click LARiver.**

4) **Name the output feature class LARiverBuffer. Because you've set the AnalysisOutputs geodatabase as the scratch workspace, the output will be saved to this location by default.**

5) **In the Distance text box, type 0.5. Click the drop-down arrow for linear units and click Miles.**

6) **Compare your settings to the figure and click Run.**

When the tool is finished running, a message at the bottom of the Geoprocessing pane notifies you that the Buffer tool completed successfully, and the new LARiverBuffer layer is added to the map.

Also, note that a warning icon ⚠ appears next to the Output Feature Class text box. Hovering over the icon notifies you that the feature class exists so that it will be overwritten if you rerun the tool.

7) **Close the Geoprocessing pane.**

8) **Confirm the presence of the new feature class by expanding the Databases folder in the Project pane and looking in AnalysisOutputs.gdb.**

Essential GIS analysis tools

Analysis tools

The tools shown here are by no means a complete list, but they include several of the ones used most often in GIS analysis. It's hard (and probably unnecessary) to give an exact definition of what makes a tool an "analysis" tool. GIS practitioners solve problems of many different kinds, and most of these problems have facets that involve spatial relationships among geographic objects: how far A is from B, how many A's are close to B, which areas are common to A and B, how you get from A to B, and so on. Analysis tools quantify these relationships among features and their attributes.

Query tools

Query tools answer questions of the form "Which features meet such-and-such condition?" Select By Attributes selects features according to an attribute value or combination of values. Select By Location selects features according to their spatial relationship to features in another layer (or sometimes the same layer). Spatial relationships include intersection, containment, adjacency, and distance.

Query tools can be accessed on the Map tab, in the Selection group, or as geoprocessing tools (Select Layer By Attribute and Select Layer By Location).

Proximity tools

The Buffer tool creates a feature class of polygons at a specified distance around input features. It is often used to draw an exclusion zone around features (no A's should be allowed within a mile of B) or to define an area of interest (you want to look for A's only within a mile of B).

The Create Thiessen Polygons tool creates a feature class of contiguous polygons around input point features. Each polygon's shape is defined by proximity to the nearest point. Thiessen polygons can be used to define allocation areas (which areas are closer to A than to any other point, which are closer to B than to any other point, and so on).

Select By Attributes

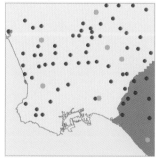

Which cities have a population of 100,000 or more?

Buffer

Streams have quarter-mile buffers.

Select By Location

Which parks are within a mile of the river?

Create Thiessen Polygons

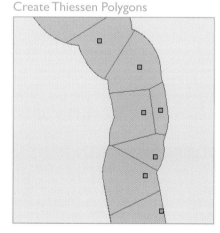

Input points are parks within a river buffer. Each output polygon encloses the area that is closer to the point it contains than to any other point.

Essential GIS analysis tools (continued)

Proximity tools (continued)

The Near tool finds the nearest feature in one or more specified layers to each feature in the input layer. It writes the distance as a new attribute to the input layer table.

Near

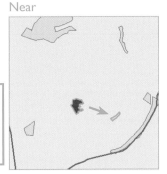

The input features are vacant parcels (symbolized in red). The features to find are parks (green). New attributes in the vacant parcels table specify the nearest park to each vacant parcel and its distance.

OBJECTID *	Shape *	NEAR_FID	NEAR_DIST
1	Polygon	117	1371.816526
2	Polygon	114	822.280138
3	Polygon	101	1234.614597
4	Polygon	34	2465.358086
5	Polygon	35	2012.800925

A table relate shows the attributes of the nearest park to the selected parcel.

OBJECTID *	Shape *	NAME	TYPE
117	Polygon	Greaver Oak Park	Local park or recreation area

Overlay tools

The basic purpose of overlay tools is to find common areas between different layers. Overlays answer questions about where A and B, each with unique and important attributes, overlap. In contrast to queries, which return selections of existing features, overlay tools create new features that have the attributes of both input layers.

The different overlay tools—Intersect, Union, Identity, and Erase are the most common—perform variations on the same basic process. They differ with respect to how much area from the input layers is included in the output feature class: common area only (Intersect), all area whether common or not (Union), one input layer's area only (Identity), or one input layer's area minus the common area (Erase).

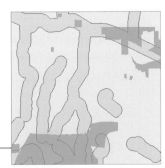

Parks and stream buffers have areas of overlap.

Intersect

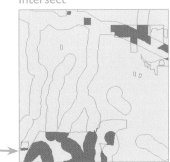

The overlay creates new features (purple) from areas of overlap. The output can be used to find park land as a percentage of riparian land or vice versa.

Spatial Join

Spatial Join is like a Select By Location query with the added benefit of joining source layer attributes to target layer features.

Spatial Join

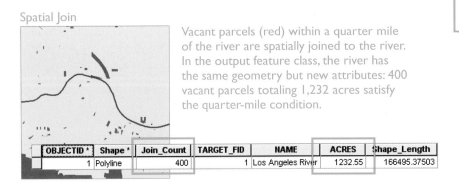

Vacant parcels (red) within a quarter mile of the river are spatially joined to the river. In the output feature class, the river has the same geometry but new attributes: 400 vacant parcels totaling 1,232 acres satisfy the quarter-mile condition.

OBJECTID *	Shape *	Join_Count	TARGET_FID	NAME	ACRES	Shape_Length
1	Polyline	400	1	Los Angeles River	1232.55	166495.37503

Buffer the parks to a quarter mile

Now you'll add the Parks data and draw a quarter-mile buffer around each park. These buffers will encompass areas that are close to existing parks and are therefore out of consideration.

1) From the LARiver geodatabase, drag Parks to the map window.

2) In the Contents pane, right-click LARiverBuffer and click Zoom To Layer.

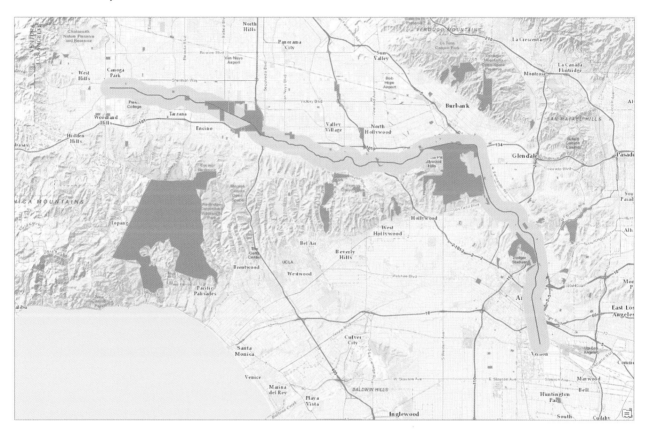

3) Turn off the LARiver layer.

You can already see a lot of overlap between existing parks and the river buffer zone. You must still extend the exclusion zone a quarter mile farther out on all sides of each park.

4) On the Analysis tab, in the Tools gallery, click Buffer.

5) Drag the Parks layer from the Contents pane into the Input Features text box of the Buffer tool.

6) **Accept the default Output Feature Class name of Parks_Buffer.**

7) **In the Distance text box, type .25. In the Linear Unit drop-down list, click Miles.**

8) **Click the Dissolve Type drop-down arrow, and then click Dissolve all output features into a single feature.**

This setting dissolves boundaries between overlapping park buffers. The dissolution of boundaries isn't strictly required, but it makes it a lot easier to interpret the result, which would otherwise be a dense tangle of lines. A side effect of dissolving the buffers is that the output will consist of one multipart feature, essentially devoid of attributes. That's okay because all you need from this layer is its geometry.

9) **Check your settings against the figure and click Run.**

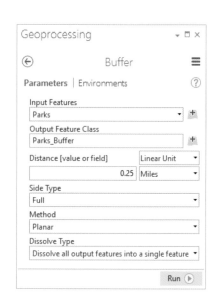

When the tool is finished running, a Parks_Buffer layer is added to the map, and a feature class is created in AnalysisOutputs.

10) **Close the Geoprocessing pane.**

11) **Turn off the Parks layer.**

12) **Change the transparency of the Parks_Buffer layer to 50%.**

▶ How? In the Contents pane, select the Parks_Buffer layer, and change the transparency using the slider on the Appearance tab, in the Effects group.

The portion of LARiverBuffer not covered by the Parks_Buffer layer represents the areas that are still under consideration. The layers are purple and tan in the figure, but your colors may be different.

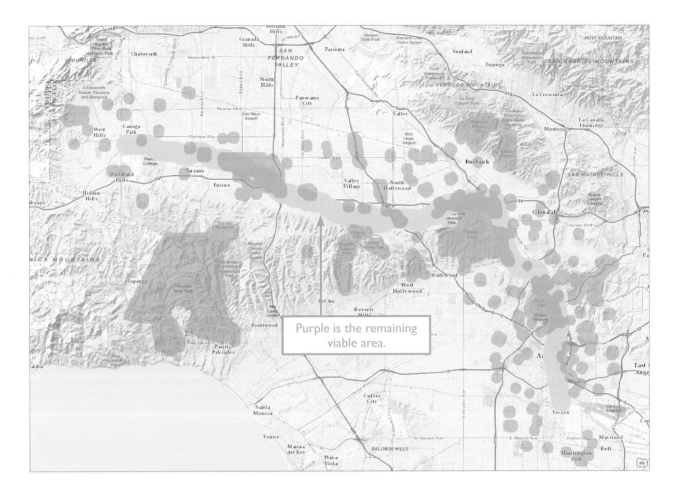

Purple is the remaining viable area.

Now you can think about what analysis you have accomplished so far. One layer, LARiverBuffer, contains all the area within a half mile of the river. You want to exclude the Parks_Buffer areas from that area, but those areas are stored in a different layer. Essentially, you want to subtract one layer from another.

Erase park-accessible areas

The tool for doing this subtraction is named Erase. This tool is a member of the family of spatial overlay operations that help make GIS such a powerful technology. Unlike the Buffer tool, the Erase tool is not available in the Tools gallery, so you'll find it using a search.

1) **On the Analysis tab, click the Tools button.**
2) **In the Search box, type Erase.**
3) **Click Erase (Analysis Tools) to open the tool.**

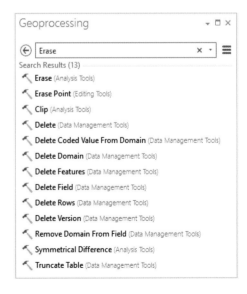

Exercise 6a: Establish proximity zones   213

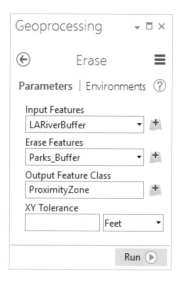

4) For the Input Features parameter, click LARiverBuffer.

5) For the Erase Features parameter, click Parks_Buffer.

6) Name the Output Feature Class parameter ProximityZone.

7) Check your settings against the figure and click Run.

When the process is finished, the ProximityZone layer is added to the map.

8) Close the Geoprocessing pane.

9) In the Contents pane, turn off all layers except ProximityZone and the basemap.

10) Symbolize ProximityZone in a medium blue (for example, Cretan Blue). This color will help the layer stand out against the basemap.

You're left with a reduced portion of the original buffer zone. Your search for a park is now confined to this shape.

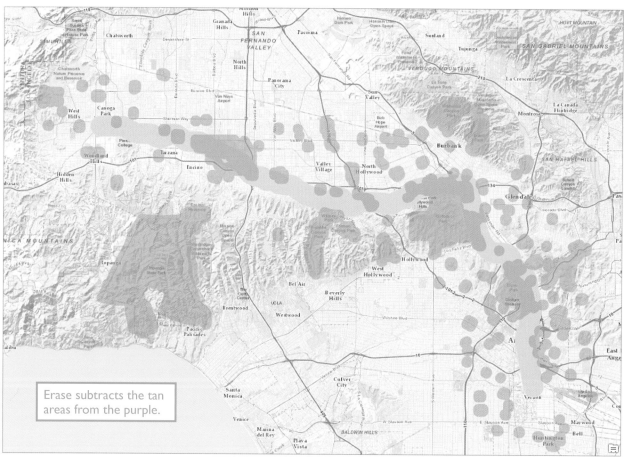

Erase subtracts the tan areas from the purple.

214   Lesson 6: Conduct the analysis

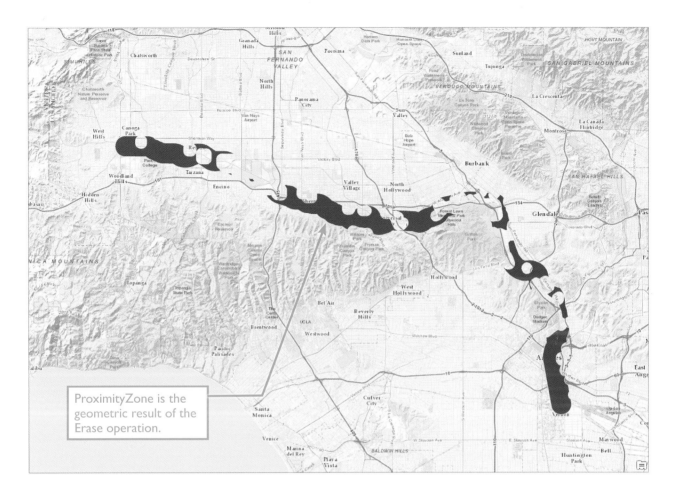

ProximityZone is the geometric result of the Erase operation.

11) Change the transparency of the ProximityZone layer to 50%. (Use the slider on the Appearance tab with the layer selected in the Contents pane.)

12) Zoom in on some of the "holes" in the ProximityZone layer.

You can see in the figure the effect that the existing parks (visible on the basemap) had on the creation of the ProximityZone layer.

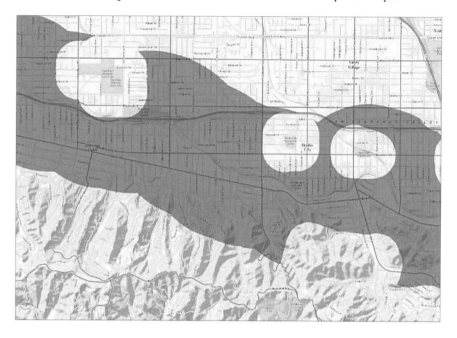

13) **Pan along the river and look at a few more examples.**

If you see some anomalies—such as a hole with no park—consider that the topographic basemap may not correspond exactly to the Parks layer. You can turn the Parks layer on and off in the Contents pane for reference.

14) **When you're finished, zoom to the ProximityZone layer.**

15) **Save your project.**

16) **If you are continuing to the next exercise, leave ArcGIS Pro open; otherwise, close ArcGIS Pro.**

In the next exercise, you'll evaluate the neighborhood demographics within the area of interest.

Exercise 6b: Apply demographic constraints

You've isolated areas that meet two of your requirements: distance to the river and distance from parks. Still ahead are the tasks of factoring in neighborhood demographics (population density, income, and age) and evaluating parcels by size and total population served. You'll deal with the demographics in this exercise, and with the parcels and population served in exercise 6c.

Within the proximity zone, you want to find areas that have the right demographic criteria. The problem is that your blue blob, if you can call it that, doesn't include those attributes; they are found only in the BlockGroups feature class. What you want to do then is make a feature class that has the spatial area of ProximityZone and the attribute values of BlockGroups.

This combination again calls for an overlay. In exercise 6a, you subtracted one layer from another, but overlay operations are typically more additive than subtractive. Usually, you want to find common ground between layers that have different attributes. The important attribute in ProximityZone is its distance to the river. The important attributes in BlockGroups are population density, age, and median income. By overlaying these two layers, you'll get all these important attributes in a single layer.

The overlay will cause splitting of any block group features that lie partly inside and partly outside the proximity zone. What happens to the attribute values of these features? If a block group with 2,000 people is halfway in the proximity zone, should the output feature defined by the overlapping half get all 2,000 of those people, or just 1,000? By default, attribute values are copied rather than apportioned in overlay operations, and copying can lead to anomalies in which large attribute values are assigned to small slivers of geometry. We've dealt with this problem ahead of time by using block group attribute values that are statistically homogenized and assumed to be uniform across the feature. It's also possible, however, to redistribute attribute values by area during geoprocessing.

Overlay block groups on the proximity zone

You want to keep all the geometry of the ProximityZone layer and only as much BlockGroups geometry as is spatially coincident with the proximity zone. This type of overlay is called Identity: it keeps all the geometry of layer A, the source layer, and only the coincident geometry of layer B, the target layer.

1) **If necessary, start ArcGIS Pro and open the LARiver_ParkSite project.**

2) **Make a copy of the Lesson6a map and rename it Lesson6b. Open the new Lesson6b map, and close the Lesson6a map by closing its tab above the map.**

3) **In the Project pane, add BlockGroups to the map from the project geodatabase.**

4) Move the BlockGroups layer under the ProximityZone layer so that you can visually see the overlap of the two layers.

5) On the Analysis tab, click Tools and search for Identity.

6) In the Tools list, click "Identity (Analysis Tools)" to open the tool.

7) For Input Features, click ProximityZone.

8) For Identity Features, click BlockGroups.

9) Change the Output Feature Class name to ProximityZone_Demographics.

10) Check your settings against the figure and click Run.

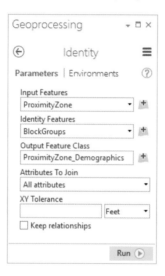

When the tool is finished running, the ProximityZone_Demographics layer is added to the map. A corresponding feature class is created in the AnalysisOutputs geodatabase.

11) **Close the Geoprocessing pane.**

12) **Zoom to the ProximityZone_Demographics layer and turn off all other layers in the Contents pane except for the basemap.**

What you see in the figure is a fractured version of the original proximity zone. ProximityZone_Demographics has the geometry of BlockGroups clipped to the boundaries of ProximityZone. As you'll see, it has the attributes of both input layers.

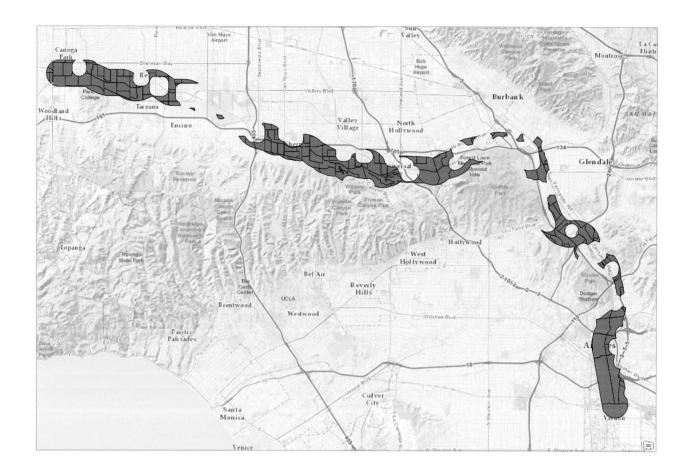

Examine attributes

Looking at the attributes of both your input and output layers will help you understand what you've accomplished by overlaying them using the Identity tool.

1) Open the ProximityZone layer attribute table—and not the ProximityZone_Demographics layer. The ProximityZone layer consists of a single record. Its NAME and BUFF_DIST (buffer distance) attributes are inherited from the LARiverBuffer layer.

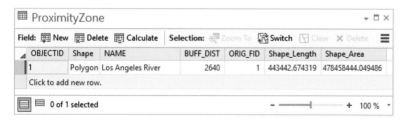

Exercise 6b: Apply demographic constraints   219

2) Open the attribute table of the BlockGroups layer. Scroll across its attributes.

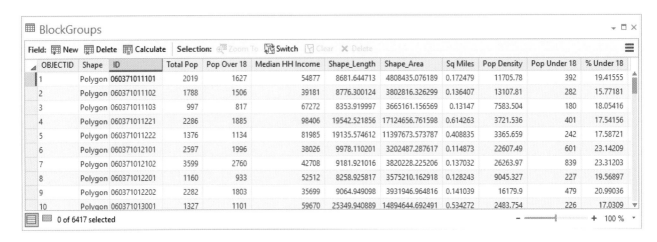

This layer has 6,417 records. It has the attributes you specified and created in lesson 4.

3) Open the attribute table of the ProximityZone_Demographics layer.

4) Drag the ProximityZone_Demographics table by its title bar, position it over the lower docking target, and release.

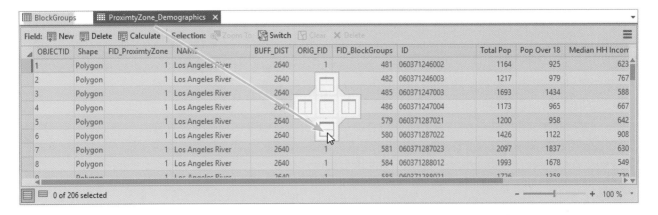

5) Resize the window so you can see both BlockGroups on the top and ProximityZone_Demographics on the bottom.

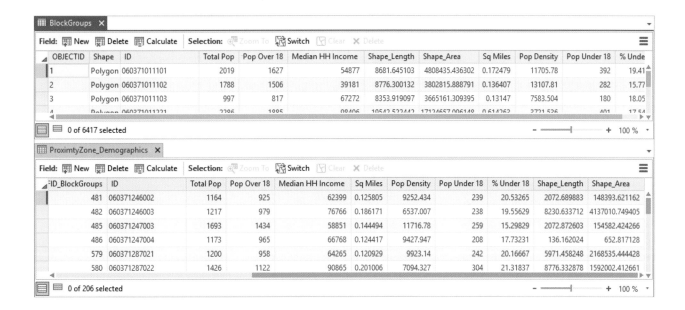

The ProximityZone_Demographics table has 206 polygons. It has all the attributes of both input layers (ProximityZone and BlockGroups), plus two ID fields that trace the features back to the input features with which they are spatially coincident.

6) **Zoom in on the western end of the ProximityZone_Demographics layer.**

▶ A map scale of around 1:24,000 is good.

7) **Turn on the BlockGroups layer.**

8) **On the Map tab, in the Selection group, click the Select tool.**

9) **Drag to draw a small box inside the feature indicated in the figure to select all features intersecting the drawn box. (A single click selects only features in the topmost layer).**

You can see in the map and in the tables that two features are selected: one in ProximityZone_Demographics and one in the BlockGroups layer underneath.

10) **Click the Show Selected Records button at the bottom of each of the tables.**

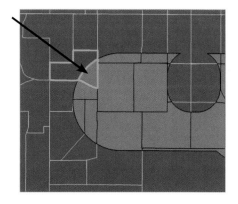

Exercise 6b: Apply demographic constraints   221

11) Scroll horizontally across the two tables so that you can see all the demographic fields.

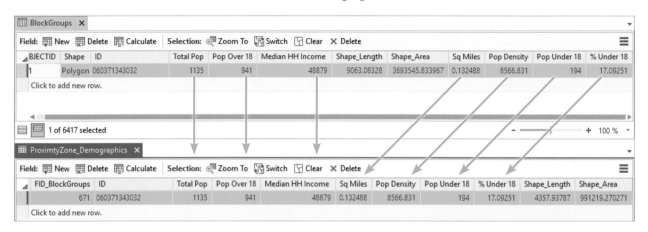

All the block group values have been copied to the spatially corresponding features in the ProximityZone_Demographics layer. This distribution of values is what you want—these cut-up block groups (some are cut up, some are intact) are the "neighborhoods" that you'll query for suitable demographics.

Note the importance of analyzing attributes such as density and percentage rather than raw numbers. The Pop Under 18 value of the selected block group is 194. You would reasonably expect this value to be smaller in the ProximityZone_Demographics feature (because it doesn't include the whole block group), but because the attribute values are copied, it's the same. It's fair to assume, however, that the % Under 18 value would remain the same even when the block group is cut into smaller pieces. This assumption is not the case, however, for the Total Pop attribute, and you might consider apportioning this value on the basis of the percentage of the area that was cut. Your analysis will not rely on apportioning because you'll be using the more detailed census block units to derive the park access population (as opposed to the block groups that you just used for the other demographic variables). For more information on apportioning, see the sidebar "Apportioning attribute values."

12) On the Map tab, in the Selection group, click Clear.

13) Close the open attribute tables.

14) Turn off the BlockGroups layer.

15) Zoom to the ProximityZone_Demographics layer.

Apportioning attribute values

Splitting feature geometry has implications for attribute values. In example 1, a census block group is split by a buffer in an overlay operation, resulting in two output features. By default, ArcGIS Pro copies the attribute values of the input feature to both output features. That's okay for POPDENSITY, which is already a ratio: the value applies to the parts as well as the whole. It's not okay for TOTALPOP, a count value, because it doubles the population, as shown.

In example 2, you want to count the population inside the buffer, which covers parts of three block groups. A reasonable approach is to analyze the percentage of each block group's area that falls inside the buffer and assign the same percentage to its population value. In this case, 1% of block group 1's area is inside the buffer, so its contribution to the buffer population should be 1% of 1,755. Likewise for block group 2 on the far right (21% of 4,445) and block group 3 in the middle (38% of 6,722).

The examples are variations of the same problem, which can be solved with a tool named Make Feature Layer. This tool makes a layer from a feature class (essentially the same thing that happens automatically when you add data to ArcGIS Pro). It has a very useful parameter, however: a "ratio policy" check box for each attribute in the input feature class. When you geoprocess a layer made with this tool, the ratio policy is applied to any split features, dividing their attribute values according to area.

The "Use Ratio Policy" box is checked for the TOTPOP_CY field.

Example 1

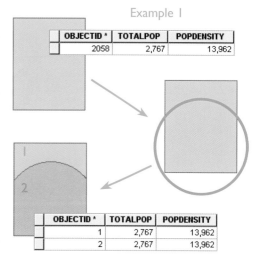

Example 2

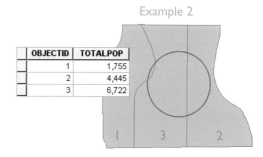

Make Feature Layer is run on the *BlockGroups* feature class. Subsequent overlay operations give the desired results.

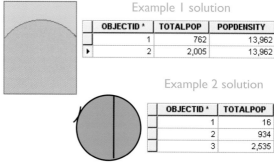

The final step to get the total population of the buffer area would be to use the Dissolve tool with TOTPOP_CY as a statistics field.

Select areas by demographic attributes

The ProximityZone_Demographics layer now contains the demographic attributes of interest: population density, the percentage of children, and median household income. You want to consider areas only if they meet your thresholds for all three values:

- population density greater than or equal to 8,500 people per square mile,
- percentage of children greater than or equal to 22 percent, and
- median household income less than or equal to $50,000.

You'll use an attribute query to select features that satisfy these conditions.

1) On the Map tab, in the Selection group, click Select By Attributes.

2) At the top of the Select Layer By Attribute tool, click the Layer drop-down arrow and click ProximityZone_Demographics. The drop-down menu defaults to the selected layer in the Contents pane so the correct layer may already be selected.

3) Make sure the Selection type is set to New selection.

4) Click Add Clause and create a clause for Pop Density is Greater Than or Equal to 8500. Then click Add.

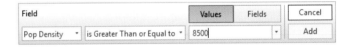

5) Click the Add Clause button again and create another clause for And % Under 18 is Greater Than or Equal to 22. Then click Add.

6) Click the Add Clause button again and create another clause for And Median HH Income is Less Than or Equal to 50000. Then click Add.

▶ If your clause is getting too long to read unhindered, hover over it and you'll see the entire thing.

Combining the queries in a single statement is more efficient than making a series of selections and subselections. The And operator ensures that features will be selected only by passing all the tests.

7) **Check your tool against the figure and click Run.**

▶ If you need to edit an expression, double-click on the clause.

8) **Close the Geoprocessing pane when the selection runs successfully and the resulting parcels are highlighted on the map.**

The figure shows the features that satisfy the query. There aren't that many, and most of them lie toward the ends of the proximity zone.

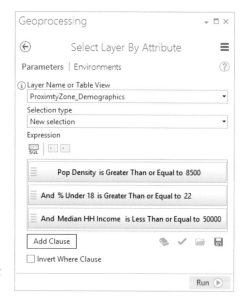

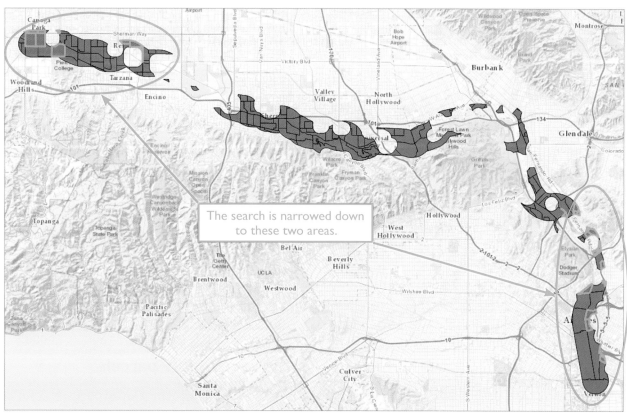

The search is narrowed down to these two areas.

9) **Open the ProximityZone_Demographics attribute table and click the Show Selected Records button.**

Exercise 6b: Apply demographic constraints   225

10) Confirm that there are 27 selected records, and review the values of the demographic fields to confirm that they meet the park criteria.

11) Close the table but keep the records selected.

Create a new feature class from the selected features

You'll save these selected features to a new feature class. Previously, in lesson 4, you used the Copy Features tool from the Tools list for this purpose. You'll use that same tool but use a more convenient method to get to it.

1) Confirm that the parcels are still selected, and then right-click ProximityZone_Demographics in the Contents pane and click Data > Export Features. This command opens the Copy Features tool with the Input Features parameter already filled out.

2) Change the Output Feature Class parameter to GoodZones.

3) Check your settings against the figure and click Run.

After the tool is finished running, the new GoodZones layer is added to the map, and a new feature class is added to the AnalysisOutputs geodatabase.

4) Close the Geoprocessing pane.

5) Turn off all layers except the new GoodZones layer and the basemap.

6) On the Map tab, in the Selection group, click the Clear button.

7) Zoom to the full extent of the GoodZones layer.

8) Save your project.

If you are continuing to the next exercise, leave ArcGIS Pro open; otherwise, close ArcGIS Pro.

Exercise 6c: Select suitable parcels

You've narrowed your focus to a layer named GoodZones. Having covered the demographic criteria, you'll turn your attention to the parcels. Thanks to your work in lesson 4, you have a dataset consisting of all, and only, vacant parcels. From this dataset, you'll select the parcels that lie within good zones. From that selection, you'll further select parcels that are one-quarter acre or larger: that selection will comprise your set of candidate park sites. Then you'll do further analysis to compare the sites by total population served and their distance to the river.

Select vacant parcels within good zones

You'll select the parcels using a spatial query.

1) If necessary, start ArcGIS Pro and open the LARiver_ParkSite project.

2) Make a copy of the Lesson6b map and rename it Lesson6c. Open the new Lesson6c map, and close the Lesson6b map by closing its tab above the map.

3) Add the VacantParcels feature class to the map from the project geodatabase.

4) On the Map tab, in the Selection group, click Select By Location.

5) Set Input Feature Layer to VacantParcels.

6) Confirm the Relationship drop-down list is set to Intersect.

7) Set the Selecting Features parameter to GoodZones.

8) Leave the Search Distance box blank.

9) Confirm that the Selection type is set to New selection.

Parcels will be selected if they lie partially or entirely within good zones.

Some parcels may cross good zone boundaries because of how the zones are clipped. Should you include parcels that straddle good zones? There's no right or wrong answer. By choosing the "intersect" spatial relationship method, you're going with a generous interpretation—you'll select a parcel if any part of it lies within a good zone because you want to maximize the candidate pool. If you wanted to be restrictive (the parcel must be entirely within a good zone), you'd choose a method such as Within.

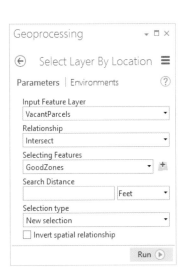

10) Compare your tool to the figure and click Run.

11) Close the Geoprocessing pane after the Selection tool runs successfully.

12) Open the VacantParcels attribute table. You should have just 55 of 14,269 parcels selected.

Select subset of parcels that are one-quarter acre or larger

Of the 55 selected parcels, you want to keep only those that are at least one-quarter acre. You'll select this subset using an attribute query on the Acres field using the features that are already selected (on the basis of the spatial query created in the last section).

1) On the Map tab, click the Select By Attributes button above the table.

2) In the Select Layer By Attribute tool, use the drop-down menu to change the Selection type to Select subset from the current selection. This setting restricts the scope of the query to records that are already selected.

3) Add a clause for:

   Acres >= 0.25.

4) Check your tool against the figure and click Run.

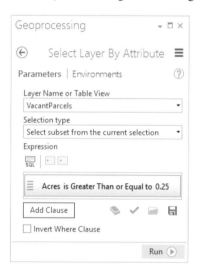

5) Close the Geoprocessing pane.

6) Look at the attribute table. Click the Show selected records button. You're down to six selected records: just six parcels meet all your conditions.

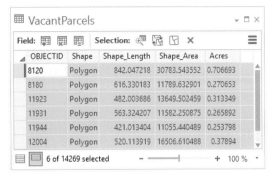

At this scale, it's hard to see the selected parcels on the map. You'll take a closer look in step 7.

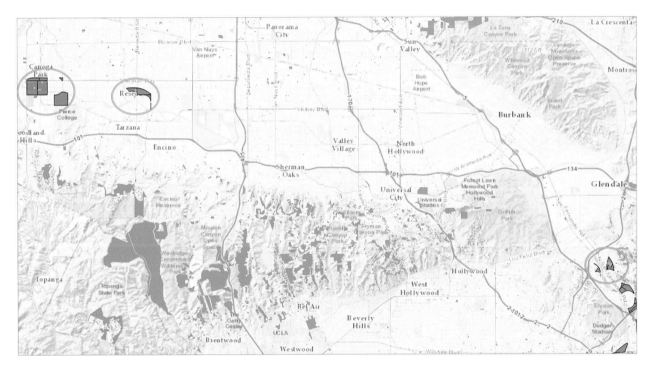

7) In the Contents pane, right-click VacantParcels and click Selection > Zoom To Selection.

8) Close the attribute table.

Export selected features

These six parcels are the critical ones you want: they represent the culmination of the process, and you want to save them as a feature class. Again, you'll use the convenient Export Data method.

1) In the Contents pane, right-click VacantParcels and click Data > Export Features.

2) In the Copy Features tool, change the Output Feature Class name to SixSites.

3) Check your settings against the figure and click Run.

After the tool is finished running, the new SixSites layer is added to the map.

4) Close the Geoprocessing pane and close any tables.

Inspect the candidates

Now you'll zoom in and take a close look at the six sites against basemap imagery.

1) On the Map tab, change to the Imagery basemap.

2) In the Contents pane, remove VacantParcels. Turn off all layers except SixSites and Imagery. The six site polygons may still be difficult to see with the current scale and symbology.

3) In the Contents pane, open the Symbology pane for the SixSites layer by clicking the color patch below the layer name.

4) In the Symbology pane, switch to the Properties view if necessary.

5) Set the fill color to No Color and the outline color to Medium Apple.

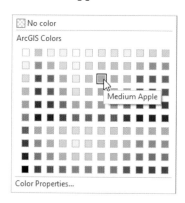

6) **Make the outline width 2.**
7) **Click Apply.**
8) **Close the Symbology pane.**
9) **Open the attribute table of the SixSites layer, and resize the table so that the map is large and all six rows are visible.**
10) **In the table, select the third record. Right-click the gray square next to the third record and click Zoom To.**

The map zooms to one of your candidate parcels. Although a portion of the parcel is being used as a parking lot, the parcel is adjacent to a quiet street to the west and could be a good park.

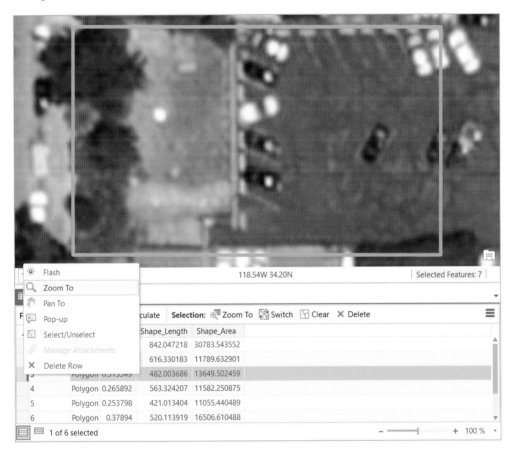

Exercise 6c: Select suitable parcels 231

11) In the drop-down scale menu below the map, change the map scale to 1:24,000.

12) If necessary, pan south so you can see the river.

13) In the Contents pane, turn on the ProximityZone layer.

14) Optionally, turn on the Parks layer and, if necessary, symbolize it in a shade of green.

The view confirms that the site is near the outer edge of the proximity zone. Even more interesting is that the site has two other parks in the immediate vicinity that are both closer to the river. That's why human interpretation of analysis results is so important. You considered proximity to existing parks, but it didn't occur to you that a park might be more than a quarter mile from a candidate site and yet closer to the river. It doesn't invalidate your candidates, but it's something to think about.

15) **Turn off the ProximityZone and Parks layers.**

16) **In the table, select the first record. Right-click the gray square next to the first record and click Zoom To.**

17) **Zoom to a map scale of 1:5,000.**

These two sites are close to the river, which is ideal. Notice a couple of other things. First, the river bottom is natural here, which could make this site more scenic than others. Your analysis doesn't take aesthetic considerations into account, but the views, sounds, and smells of a park affect the people who use it.

On the other hand, the site is also close to a freeway. This location has an aesthetic impact and probably a health impact as well should kids be exposed to all that car and truck exhaust.

Investigate the remaining sites

1) Zoom to the remaining sites and investigate them:
    - Zoom and pan to see the site in relation to the river.
    - Turn other map layers on and off.
    - Make observations that add context to the analysis.

This site is close to the river and a bike path.

This site is close to the river and on a corner parcel.

This site is adjacent to a new Metrolink stop.

234   Lesson 6: Conduct the analysis

2) When you're finished, click the Clear button to remove any selections in the table.

3) Turn off all layers except SixSites and the basemap.

▶ Leave the SixSites attribute table open. Confirm there are no selected records in the table. If necessary, click the Clear Selection button above the table.

At this point, two items remain from your list of analysis criteria: you want the new park to serve the maximum number of people, and you want it to be as close to the river as possible.

Calculate distance to the river

You know that all the sites are within a half mile of the river, but knowing the exact distance from each site would be a useful metric to help decision-makers prioritize the sites. You could use the Measure tool to do that, of course, but you'd face the same problem of repetitively writing the results to a table. Also, you might not be sure of making the shortest possible measurement. The Near tool will make these calculations for you automatically.

1) On the Analysis tab, in the Tools gallery, open the Near tool .

2) In the Near tool, set Input Features to SixSites.

3) Set the Near Features parameter to LARiver.

Note that you can add multiple layers to the list of Near features. Typically, the tool is used to discover which feature, in one or more layers, is closest to a feature of interest. In this case, you want only one distance measurement.

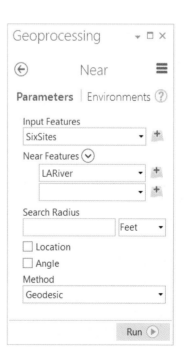

4) Using the drop-down menu, change Method from Planar to Geodesic.

This method gives you the most accurate distance, regardless of the map projection. For more complicated datasets with more features, processing using this method may take longer because of the more complex geodesic distance calculations. But again, in this case, you want only one distance.

Note that, unlike most of the other tools you've used in this lesson, the Near tool doesn't have an Output Feature Class parameter. The Near tool does not create a new feature class. Instead, it will add a couple of fields to the SixSites table.

5) Check your settings against the figure and click Run.

6) **If necessary, open the SixSites layer attribute table.**

The NEAR_DIST field stores the distance, in feet, of each site to the river. This distance unit is dependent on the coordinate system. Because your coordinate system uses feet as the linear unit, your distance units are in feet.

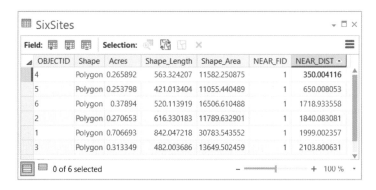

7) **Sort the attribute table by NEAR_DIST in ascending order to see how the sites rank in terms of proximity to the river.**

8) **Close the table.**

Add block centroids to the map and visualize the analysis

You must find the population within a quarter mile of each potential park site. The basic approach of counting the block centroids around each park and summing their populations was introduced in lesson 2 and sketched in the analysis plan at the beginning of this lesson.

You can do that using the Spatial Join tool. You might think of a spatial join as a spatial query plus a table join: attributes from one table are joined to another on the basis of a spatial relationship between layers rather than possession of a common attribute.

Now you can get a visual sense of what you'll accomplish with the spatial join by examining the population data within a quarter mile of a park site.

1) **Zoom in on one of the sites.**

2) **Change the scale to 1:10,000 using the drop-down scale menu below the map.**

3) Add BlockCentroids to the map from the project geodatabase. With the imagery basemap, you can clearly see the blocks that the point centroids represent.

4) Open the BlockCentroids attribute table and find the POP2010 attribute. This is the attribute you'll use to find the population within a quarter mile.

5) Turn off all layers (including the imagery basemap) except BlockCentroids and SixSites.

6) On the Map tab, in the Inquiry group, click the Measure tool. Confirm that Measure Distance is selected by clicking the drop-down arrow under the button.

7) Change the distance units to miles by clicking Options from the Measure panel in the upper-left corner of the map.

8) Click the center of the site and move your pointer to measure out 0.25 miles. Double-click to stop measuring.

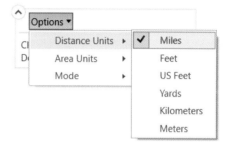

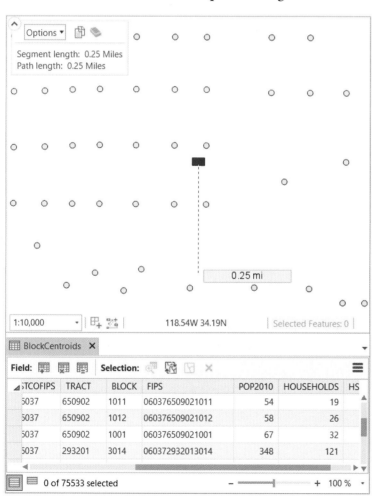

Exercise 6c: Select suitable parcels   237

In the next section, you'll use the POP2010 field values and sum the values within this quarter-mile radius as a simple means of determining how many people are within easy walking distance of the park.

9) **Change the active Measure tool back to Explore by clicking the Explore button on the Map tab, in the Navigate group.**

10) **Close the BlockCentroids table.**

Determine population around the park site

In lesson 2, you decided to sum the population in terms of a quarter-mile distance from the park site. This amount is an oversimplification of accessibility, but it has the virtue of being easy to analyze. You'll use block centroids for this measurement, because they are the most detailed geographic unit maintained by the census. Because of their small size, the block centroid population values are not as easily estimated and are therefore only updated with the decennial census every 10 years. This method will be acceptable for this analysis but is definitely something to consider for fast-growing areas.

1) **Right-click the SixSites layer and click Zoom To Layer. You probably can't see the sites anymore, but that's okay.**

2) **Clear any selections by clicking the Clear button on the Map tab, if necessary.**

3) **On the Analysis tab in the Tools gallery, click Spatial Join.**

4) **In the Geoprocessing pane, set Target Features to SixSites.**

5) **Set Join Features to BlockCentroids.**

6) **Change the Output Feature Class name to SixSites_AccessPop.**

7) **Leave the default Join Operation parameter as Join one to one.**

The Field Map of Join Features area defines which fields will be in the output. The only fields you need in the output are the ACRES, NEAR_DIST, and POP2010 fields so you'll remove all the others. You can also do some cleanup of field names at this point so you'll change the name of the NEAR_DIST attribute to something more descriptive.

8) In the Output Fields list in the Field Map of Join Features area, click Shape_Length, and then click the red X that appears on the right of the field name.

9) Repeat step 8 for all other fields in the list except ACRES, NEAR_DIST, and POP2010.

▸ If you remove too many fields by mistake, you can reset the Output Fields list by clicking the Reset button 🔄 in the upper-right corner of the list.

10) In the Output Fields column, click POP2010, and then click Sum in the Merge Rule drop-down list, in the Source area on the right.

The Merge Rule setting is crucial. It tells ArcGIS Pro that for each target feature (park site), you want to sum the population values of the join features (block centroids). Note that other merge rules are available to select.

11) Click the Properties tab (just above the Merge Rule drop-down arrow), and change both Field Name and Alias to AccessPopulation.

12) In the Output Fields column, click NEAR_DIST, and change both Field Name and Alias to RiverDistance.

13) In the Match Option drop-down box, click Within a distance.

14) Enter 0.25 for Search Radius, and confirm that the distance is set to miles.

15) Compare your tool to the figure and click Run.

When the tool is finished running, a SixSites_AccessPop layer is added to the map.

16) Close the Geoprocessing pane.

17) Open the attribute table of the SixSites_AccessPop layer.

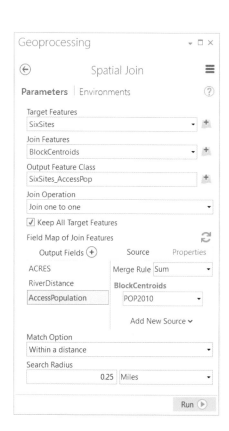

The AccessPopulation field has the total population within a quarter mile of each site. You've also taken the opportunity to clean up the table so it's easier to interpret the results.

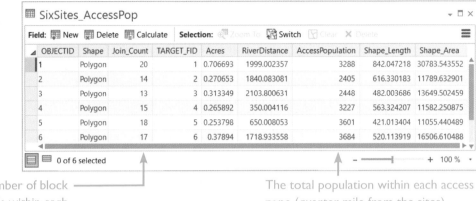

The number of block centroids within each access zone.

The total population within each access zone (quarter mile from the sites).

18) **Close the table.**

Add the demographic attributes to SixSites_AccessPop

It would be nice to store all the relevant site-selection attributes in a single table with the candidate sites. These attributes include:

- Acreage
- Distance to the river
- Total population within a quarter mile
- Demographic variables

Doing so will make it easy to evaluate and compare the sites at a glance—for you and for anyone you share the data with. You already have the acreage, distance, and population attributes in the SixSites_AccessPop table. In this section, you'll use the Identity tool to get the demographic attributes into the table. These attributes are found in the BlockGroups layer.

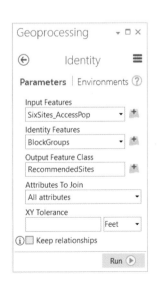

1) **Open the Identity tool (browse or search from the Geoprocessing pane).**

2) **For Input Features, click SixSites_AccessPop.**

3) **For Identity Features, click BlockGroups.**

4) **Rename Output Feature Class as** RecommendedSites.

5) **Check your settings against the figure and click Run.**

When the tool is finished running, the RecommendedSites layer is added to the map.

6) Close the Geoprocessing pane.

7) Open the RecommendedSites attribute table and scroll across it.

It has all the attributes of interest, plus several attributes that you don't need along with some ID attributes (back links to features in other tables) that you don't need for this exercise. In the next exercise, you'll format this table to make it presentable.

8) Close the table.

9) In the Project pane, confirm that your AnalysisOutputs geodatabase looks like the figure.

10) Save your project.

11) If you are continuing to the next exercise, leave ArcGIS Pro open; otherwise, close ArcGIS Pro.

- Databases
  - LARiver_ParkSite.gdb
  - AnalysisOutputs .gdb
    - GoodZones
    - LARiverBuffer
    - Parks_Buffer
    - ProximityZone
    - ProximityZone_Demographics
    - RecommendedSites
    - SixSites
    - SixSites_AccessPop

Exercise 6d: Clean up the map and geodatabase

Analysis projects tend to leave clutter behind. It's normal to find your working map and geodatabase filled with a mixture of data you want to preserve and data you can now discard. It's tempting to leave the mess behind, but sooner or later someone will return to the scene and wish that you'd taken the time to put things in order. In this exercise, you'll simplify your map, which contains several unnecessary layers. You'll also format the SixSites attribute table and save it as a layer file. Finally, you'll remove intermediate data from the geodatabase and add metadata to the datasets you want to keep.

Clean up the map

This map isn't the one you'll use for your final layout, but you still want it to be intelligible. With that in mind, you'll keep just a few important layers in it.

1) If necessary, start ArcGIS Pro and open the LARiver_ParkSite project.

2) Make a copy of the Lesson6c map and rename it Lesson6d. Open the new Lesson6d map, and close the Lesson6c map by closing its tab above the map.

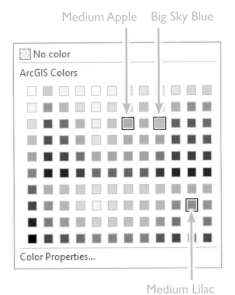

Medium Apple   Big Sky Blue

Medium Lilac

3) In the Contents pane, remove all layers except the four listed as follows. Put the layers in the order shown, from top to bottom:
    - LARiver
    - RecommendedSites
    - LARiverBuffer
    - World Imagery (the basemap)

4) Turn all four layers on and zoom to the LARiverBuffer layer.

5) If necessary, symbolize RecommendedSites with a fill color of No Color, an outline color of Medium Apple, and an outline width of 2.

6) Symbolize the LARiver layer with the color Big Sky Blue and a width of 3.

7) Symbolize LARiverBuffer with a fill color of No Color, an outline color of Medium Lilac, and an outline width of 3.

Your map should look like the figure.

Add layer descriptions

When you add a layer to a map, a description—taken from the item description of the source data—is added to the layer properties. This description explains what the layer represents. Layer descriptions are helpful to anyone using the map and are required if you share layers as layer packages. The LARiver layer already has a description, as does the World Imagery sublayer of the basemap; the other two layers, created during geoprocessing, do not.

1) **Open the layer properties of RecommendedSites (right-click the layer) and click the Metadata tab on the upper left.**

2) **In the Description box, type** These are the six sites recommended for a new park near the Los Angeles River. **Click OK.**

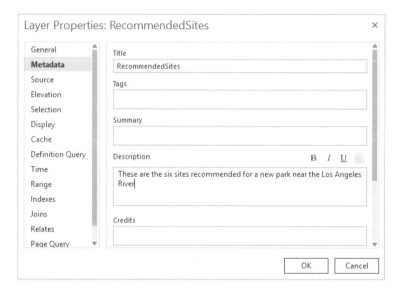

3) **Open the layer properties of LARiverBuffer. In the Description box, type** This is a half-mile buffer around the Los Angeles River. **Click OK.**

Delete unnecessary fields

In exercise 6c, you got all the important attributes into the RecommendedSites table. You also have several attributes you don't need, along with several FID attributes that you don't need to maintain.

1) **Open the RecommendedSites attribute table.**

2) In the table, right-click the FID_SixSites_AccessPop field heading and click Delete.

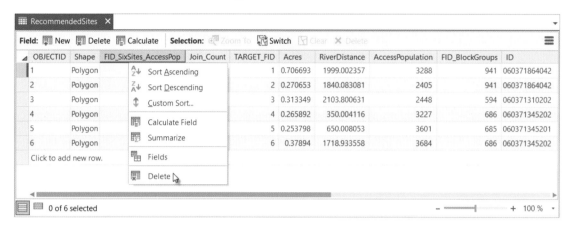

3) Click Yes on the prompt to confirm the field deletion.

The field is deleted from the table.

▶ Be careful when you delete a field because you can't undo it.

4) In the same way, delete all the remaining unnecessary fields from the table:
   - Join_Count
   - TARGET_FID
   - FID_BlockGroups
   - ID
   - Total Pop (this was the block group's population, not the more detailed block sum)
   - Pop Over 18
   - Sq Miles
   - Pop Under 18 (you need only the percentage)

The table should look like the figure. If you have more fields than those listed, go ahead and delete them.

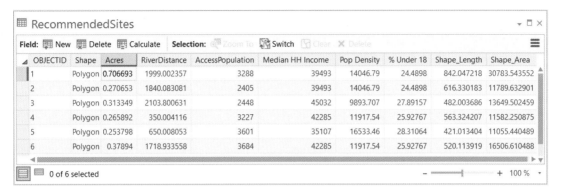

Add a SiteID field

You're going to add an identifier field that you can manage. You'll have a reason to use this field in lesson 8.

1) Click the Add Field button at the top of the table.

2) Name the field SiteID, give it an alias of SiteID, and change Data Type to Short.

3) Check your settings against the figure, and save your field changes by clicking Save on the Fields tab, in the Changes group.

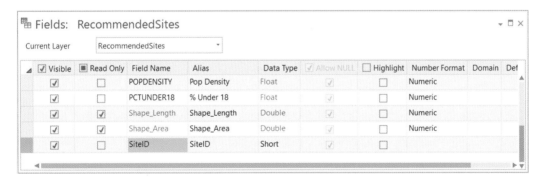

4) Close the fields view to return to the attribute table.

5) Scroll horizontally to the end of the table. Right-click the SiteID field heading and click Calculate Field.

6) In the Calculate Field tool, double-click OBJECTID in the list of fields to add it to the Expression box.

Exercise 6d: Clean up the map and geodatabase

This expression will assign the same values of 1 to 6 to the new field. The difference is that SiteID is a field you can manage, whereas OBJECTID is not.

7) **Check your settings against the figure and click Run.**

The field is populated with the values from the OBJECTID field.

8) **Close the Geoprocessing pane.**

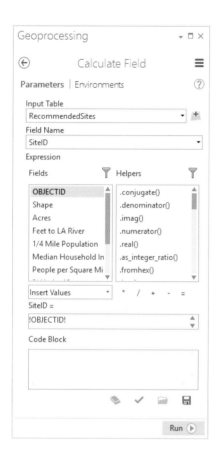

Format fields

Most of the fields in the table should be given an alias or formatted in other ways to make them more readable. We'll go through the steps in detail for one field and let you do the rest with general instructions.

1) **Open the fields view for RecommendedSites by clicking the Menu button on the right of the table and then clicking Fields View.**

2) **Confirm that the Current Layer drop-down box above the table is displaying RecommendedSites (Lesson6d).**

3) **In the list of fields, in the Visible column, click to clear Shape_Length and Shape_Area to hide these fields. On the Fields tab, click Save.**

4) **At the bottom of the list, press and hold the gray square on the left of the SiteID row and drag it to the top of the list just below the Shape field.**

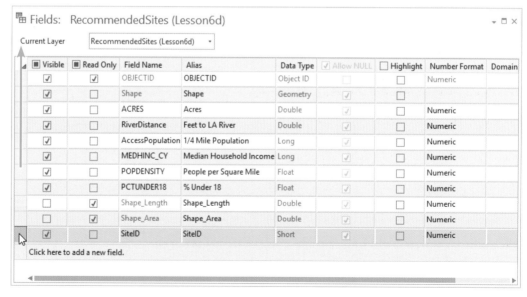

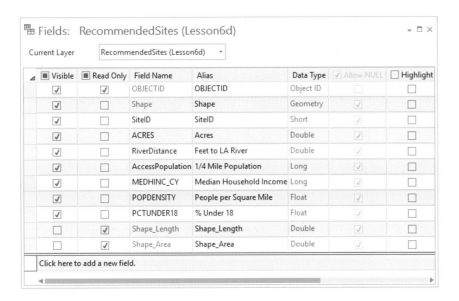

Moving this field affects only this layer. In the feature class, fields will continue to be stored in the order in which they're added. If you drag this feature class to this map or any other, the fields will be displayed in their original order.

5) **Find the row for the PCTUNDER18 field and scroll to the right to find the Number format column.**

6) **Double-click in the cell that currently contains the text Numeric and click the ellipse button on the right of the cell.**

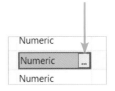

The Number Format dialog box appears to more specifically define how the attribute values should be displayed in the map.

7) **In the Category drop-down list, click Percentage.**

8) **Confirm that Number already represents a percentage is selected.**

9) **Change the number of decimal places to 0.**

10) **Compare to the figure and click OK.**

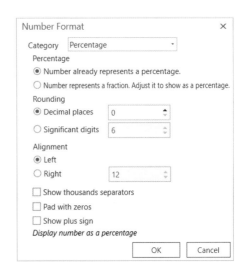

Exercise 6d: Clean up the map and geodatabase   247

11) Format the remaining fields per table 6-1. Some of these fields may already be correct in your table, but depending on the field, you may have to
   - Change or add an alias
   - Change the number format
   - Change the number of decimal places
   - Select the check box to show thousands separators

Table 6-1. Field changes

Field	Alias	Number Format
ACRES	Acres	Numeric (one decimal place)
RiverDistance	Feet to LA River	Numeric (zero decimal places; show thousands separators)
AccessPopulation	1/4 Mile Population	Numeric (show thousands separators)
MEDHINC	Median Household Income	Currency
POPDENSITY	People per Square Mile	Numeric (zero decimal places; show thousands separators)

12) On the Fields tab, click Save.

13) Close the fields view to return to the attribute table. The attribute table should now look like the figure.

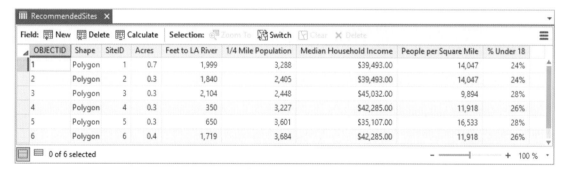

14) Close the attribute table.

Save the layer as a layer file

All these format settings are layer properties. If you want to make them available in other maps, you must save RecommendedSites as a layer file, as you did with the city boundary of Los Angeles in lesson 1.

1) In the Contents pane, right-click RecommendedSites and click Save As Layer File.

2) In the Save Layer(s) As LYRX File dialog box, navigate, if necessary, to the MapAndMore folder.

3) Name the file RecommendedSites.

Exercise 6e: Evaluate your results

The recommended sites meet your criteria and look good on basemap imagery, but ultimately, there's no substitute for on-site inspection. How do these candidate locations really look in the context of their surroundings? If you had the opportunity, it would be worthwhile to visit the sites and record your impressions. In this exercise, we've included some photographs of the sites to help give you a sense of that experience.

Compare your guesses from lesson 1

In lesson 1, you made some guesses about likely areas for parks on the basis of a preliminary look at the data. Now you can see how those guesses turned out.

1) If necessary, start ArcGIS Pro and open the LARiver_ParkSite project.

2) Make a copy of the Lesson6d map and rename it Lesson6e. Open the new Lesson6e map, and close the Lesson6d map by closing its tab above the map.

3) Open the Lesson1b map by double-clicking it in the Project pane, under Maps.

4) Drag the Lesson6e tab to the right-hand docking target that appears in the middle of the map.

The Lesson6e map is placed on the right of the Lesson1b map so you can compare them side by side.

5) **On the View tab, click the drop-down arrow below the Link Views button and click Center And Scale. A link icon appears on both map tabs to show that the map views are linked.**

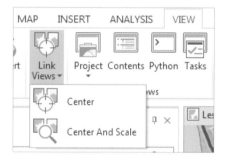

The two maps are linked, so that as you navigate one, the other map's center and scale will be synchronized. Linking views links the views of all the open maps so this same method can be used to compare as many maps (and 3D scenes, which display data on a globe) as you want.

6) Click the Lesson1b tab to make it the active map. The active map determines which layers are displayed in the Contents pane.

7) Zoom to the Los Angeles River layer. Both maps zoom out to the length of the LA River line.

Copy layers between maps

Another way to make this comparison is to add the layer that has your guesses (the Map Notes you created in the Lesson1b map) to the Lesson6e map.

1) If necessary, make Lesson1b the active map by clicking its tab above the map.

2) Right-click Bright Map Notes and click Copy.

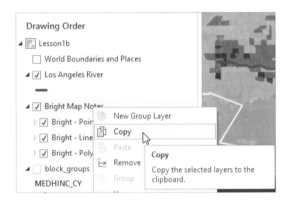

3) Activate the Lesson6e map by clicking its tab above the map.

4) In the Contents pane, right-click Lesson6e and click Paste. The Bright Map Notes group layer (containing point, line, and polygon sublayers) is added to the map.

5) On the View tab, in the Link group, click the Link Views button again to turn it off. The link icon will disappear on both map tabs.

6) Close the Lesson1b map.

7) Zoom and pan around the Lesson6e map to see how your guesses (purple stars) compare with the analysis results.

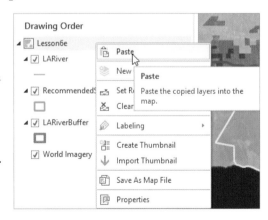

Examine park candidates

You'll take a short virtual tour of the six candidate sites and create bookmarks of each site.

1) **Open the RecommendedSites attribute table and zoom to Site 1. Create a new bookmark, Site 1, on the Map tab, under Bookmarks.**

Ground-level images of Site 1.

2) **Zoom to Site 2 and create a Site 2 bookmark.**

Ground-level images of Site 2.

3) **Zoom out until you see both Sites 1 and 2.** As noted earlier, both sites are close to the river, which is ideal. Both sites have steep slopes but do have good access.

4) **Zoom to Site 3 and create a Site 3 bookmark.**

Site 3 is on a quiet street and is level, but part of the site is being used as a parking lot. This would require some additional work to create a park at this location.

Ground-level images of Site 3.

5) **Zoom to Site 4 and create a Site 4 bookmark.**

Ground-level images of Site 4.

Ground-leve

Ground-leve

Lesson 7 Automate the analysis

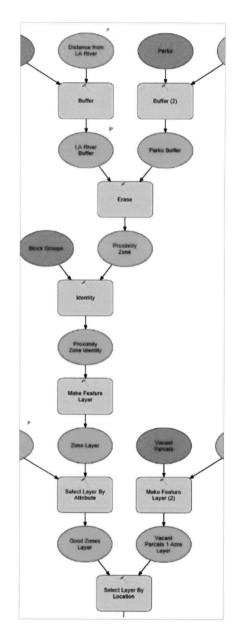

SUPPOSE UNEXPECTED CHANGES

occur after you've finished the analysis. Maybe an updated table of vacant parcels becomes available. Maybe after seeing the results, the city council wants to add or remove criteria, or see how a change in threshold values affects the outcome. Of course, you could redo the analysis, but even a small change—especially if it comes at the beginning of the process—could make a lot of tedious work. This is the moment to introduce ModelBuilder™. ModelBuilder can encapsulate your entire workflow in a model—a custom geoprocessing tool that runs with the push of a button.

A model looks like a flowchart or schematic diagram that shows the steps for manufacturing a product. Inputs, outputs, and processes are represented by geometrical shapes. Arrows connect the shapes to show the sequence of operations. For an example, see the sidebar "Understanding a model" later in this lesson.

A model is more than a visual aid, it's also an engine. The datasets and tools it represents are embedded in the model's schematic elements, so that running a model sets in motion a chain of geoprocessing operations. In this case, the model will basically repeat the analysis tasks you did in lesson 6.

There are lots of good reasons to build a model from your analysis. First, since it is a diagram, a model makes the whole workflow transparent and open to scrutiny. This visualization may reveal the analysis's weaknesses or confirm its validity—either way, it brings the methodology into the open. Second, a model stores, and thereby documents, the workflow, sparing you the task of keeping detailed notes on what you did and why you did it. Third, models are flexible. They easily accommodate the kinds of changes to inputs or parameters mentioned above with little effort on your part. Fourth, models can be shared with colleagues, saving development work and encouraging collaboration on sound methods and best practices. Fifth, models run quickly—just wait and see.

Lesson Seven road map

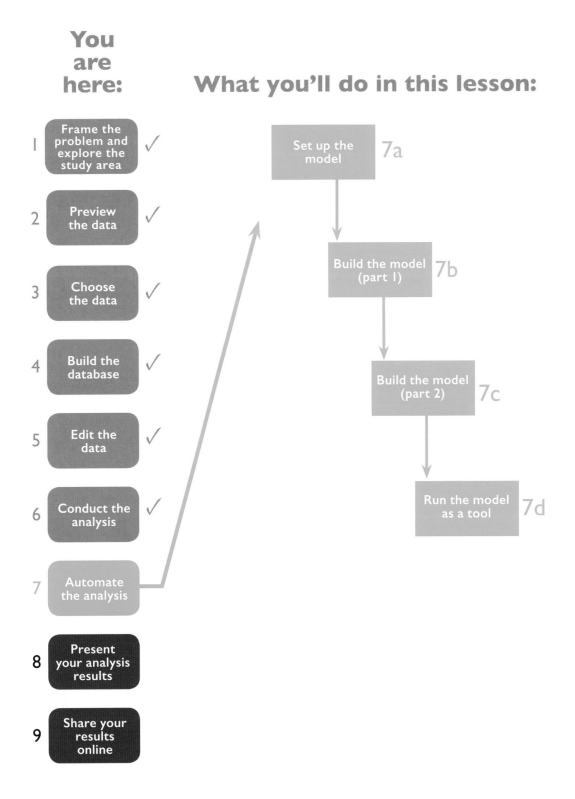

Exercise 7a: Set up the model

Create an output geodatabase and set your workspaces

As in lesson 6, you'll create a new geodatabase to hold your output data. You'll keep the results in a separate geodatabase simply to avoid confusion between the outputs that were created in lesson 6.

1) Start ArcGIS Pro and open the LARiver_ParkSite project.

2) Insert a new map and name it Lesson7. Close any other maps you may have open.

3) In the Project pane, right-click Databases and click New File Geodatabase. A dialog box will open to save the new geodatabase.

4) Name the new geodatabase ModelOutputs and click Save.

5) On the Analysis tab, in the Geoprocessing group, click the Environments button.

6) To set the current workspace, click the browse button next to the Current Workspace box and browse to your new ModelOutputs geodatabase.

7) Set Scratch Workspace to ModelOutputs as well.

8) Compare to the figure and click OK on the Environments dialog box.

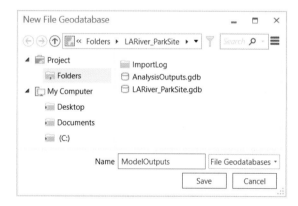

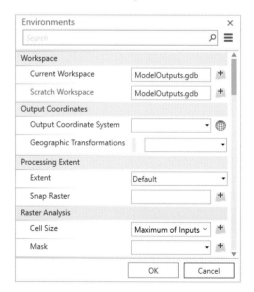

Exercise 7a: Set up the model   259

As a reminder, these "environment settings" set the default output location for geoprocessing tools. You can always change the output location for any individual tool, but setting these default workspaces is convenient when trying to keep your outputs in the same geodatabase.

Create a model in the project toolbox

A toolbox is a container of tools. A model is a type of tool (just like Buffer, Identity, and the other tools you've worked with) and, as such, must go in a toolbox. However, it can't go in a system toolbox (like the Analysis toolbox you worked with in lesson 6). An ArcGIS Pro project automatically contains a project toolbox, so when you created the ArcGIS Pro project in lesson 1, the project toolbox was created for you.

1) **In the Project pane, expand Toolboxes. Here you'll find a toolbox with the same name as your project.**

2) **Expand the toolbox, and notice that it is empty.**

3) **Right-click the project toolbox and click New > Model.**

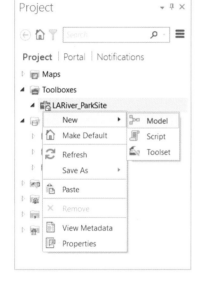

An empty Model window opens, usually replacing the map view. If it doesn't open to replace the map view, drag the Model tab to the center of the map and dock it. You can always return to the map by clicking the map tab at the top of the pane.

Also note that a new tool named Model is in your toolbox, and it has a model icon next to it signifying that the type of tool is a model. Models are a special kind of tool, but you'll see that, once completed, they can behave like any other tool in the toolbox.

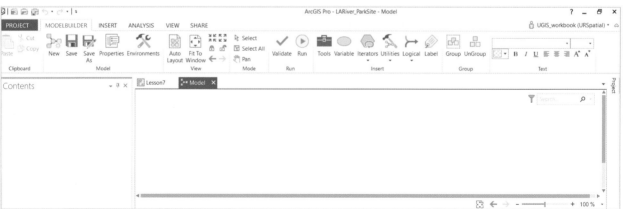

260   Lesson 7: Automate the analysis

Understanding a model

To get the basic idea, forget about the LA River for a moment and prepare a Saturday morning breakfast. On the right is a recipe (workflow) for making a dozen classic American pancakes from scratch. You should pick up the ingredients at your local store and make a batch to celebrate when you have completed this workbook!

The ingredients are symbolized by blue ellipses (inputs). These are prepared in various ways as shown by the orange rectangles (tools or processes). The results are green ellipses (outputs). You stir the basic ingredients to make the flour mix before adding the whipped egg, oil, and buttermilk. The mix should be properly blended before liquids are added; otherwise, you may get clumps of baking powder in the dough, which will lead to some very flat pancakes. After you make the dough, refrigerate it for several hours (or, even better, overnight) to allow time for chemical reaction. Then fry the dough in a pan with your favorite fat and eat it with a good brand of maple syrup. Delicious!

Can anything go wrong? If you follow the arrows, you're on a tried and true path. If you start improvising, you might make the recipe even better—but you might also get your weekend off to an indigestible start.

It's much the same with an ArcGIS workflow. With a complex analysis, it usually takes trial and error to get the recipe right. Typically, you'll experiment with several combinations of data, tools, and sequences before finding the best way to solve your problem. Once you find it, you'll want to save it and share it with friends, just like the pancake recipe.

We experimented a lot with the data and many ArcGIS tools before writing this cookbook (sorry, workbook). Having figured out the workflow, we implemented it in lesson 6. But we also wanted to preserve it, encapsulate it, and make it a formula—as you'll see, that's what ModelBuilder is all about.

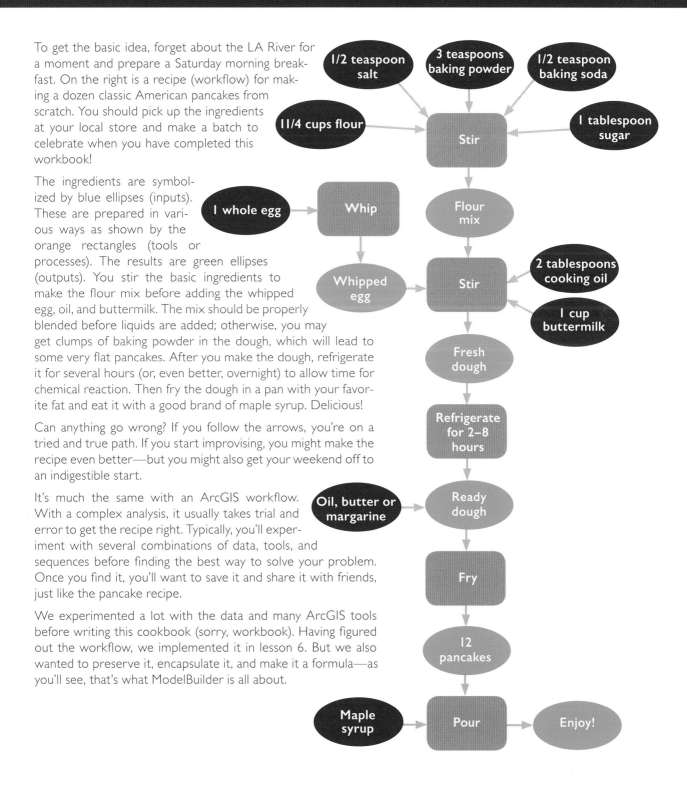

Exercise 7a: Set up the model   261

Set model properties

You'll rename the model and supply some information about it.

1) With the empty model displayed, on the ModelBuilder tab click the Properties button . A dialog box opens for some basic information about the model.

2) In the Name text box, type ParkSuitabilityAnalysis (no spaces).

3) In the Label text box, type Park Suitability Analysis (using spaces this time).

The label is like a field alias: a name that's more descriptive or easier to read that will be used while you're in ArcGIS Pro.

4) Verify that the box is checked for Store tool with relative path.

The model, its input data, and its output data can now be moved without disruption as long as they remain in the same workspaces (folders or geodatabases) relative to each other.

5) Compare to the figure and click OK.

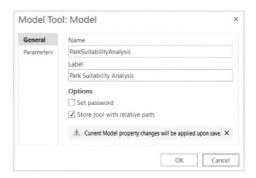

Note that the name of the model on the tab above the model canvas has updated, but the model in the Project pane hasn't. You must save the model to commit any changes made to the model.

6) On the ModelBuilder tab, in the Model group, click Save .

In the Project pane, the model now appears with its new label.

Exercise 7b: Build the model (part 1)

Your model will re-create the analysis workflow from lesson 6. Knowing how much work that lesson entailed should give you insight into why models are such an efficient tool: not only for analysis projects, but for any series of operations that you might want to save, repeat, vary, or use as a basis for further development. Re-creating the analysis will let you focus your attention on the modeling process itself, since the data and geoprocessing tools will be familiar.

The data preparation tasks from lesson 4 won't be part of the model. A big advantage of models is that it's easy to rerun them. You don't want to incorporate operations such as adding fields, calculating field values, joining tables, and so on, which only need to be done once.

In this exercise, you'll model about half the analysis steps, saving the remainder for the next exercise. The stopping point is more or less arbitrary, but building a model incrementally is good practice. It's easier to find and fix problems if you do test runs along the way.

Add data to ModelBuilder

In this section, you'll add all the input data for the model. This is the opposite of what you did in lesson 6, in which you added layers one at a time as you needed them. Either way is fine: it's a matter of preference.

1) In the Project pane, expand the project database.

2) Press and hold the Ctrl key and click these five feature classes:
   - BlockCentroids
   - BlockGroups
   - LARiver
   - Parks
   - VacantParcels

3) With the feature classes selected, drag them from the Project pane to the model (as you would drag them to a map).

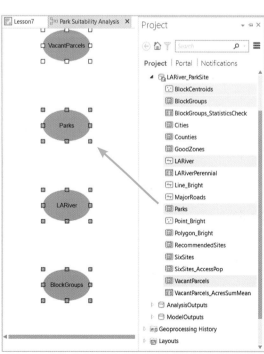

Although you can probably see only one or two of them without scrolling, each dataset is added as a blue ellipse to the Model window. (These blue ellipses are "input data variables," but you'll mostly just call them *inputs*.) The elements aren't automatically sized to fit so you'll take care of that next.

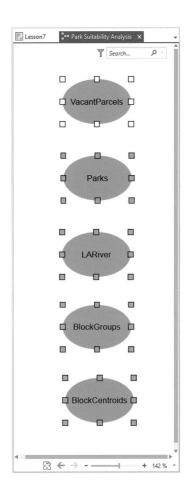

4) On the ModelBuilder tab, in the View group, click the Fit To Window button.

You should see all five elements centered in the Model window and marked with gray selection handles.

The display of text within ellipses changes as you zoom in or out in the Model window. The places where words divide may change and so may the number of lines of text. Don't worry if the text display on your screen is different from the lesson figures.

5) Hover over any selected element. Then drag the whole group to the left side of the Model window.

6) Click in empty white space in the Model window to deselect the elements.

Add a tool to ModelBuilder

The first step of the analysis in lesson 6 was to buffer the LA River. So you can start with that.

1) On the ModelBuilder tab, in the Insert group, click the Tools button. The Geoprocessing pane displays the list of toolboxes, which should be familiar from lesson 6.

2) Switch the view to Toolboxes (tab at the top of the pane), and browse to the Buffer tool: Analysis Tools > Proximity > Buffer.

3) Drag the Buffer tool from the Geoprocessing pane and drop it on the model.

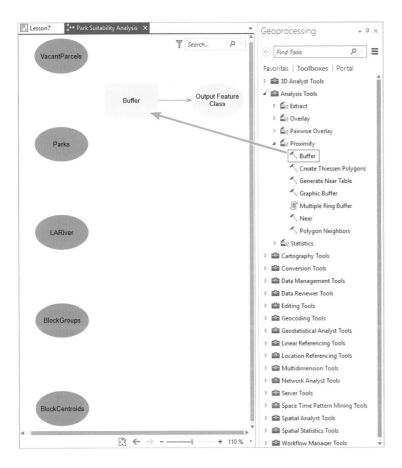

The tool is added to the Model window, connected by a small arrow to an Output Feature Class element. The tool and its output are gray because they're not ready to run. The tool must be connected to input data and its parameters be filled out.

4) **Make sure the Buffer tool and Output Feature Class elements are selected (marked with gray selection handles).**

▶ If they're not, using the Select tool in the Mode group, drag to draw a box around these elements to select them.

5) **Drag the selected elements down and place them on the right of the LARiver input data element.**

The exact position doesn't matter.

6) **Place the pointer over the blue LARiver input data element until the hand cursor 👆 appears.**

7) With the hand cursor displayed, drag a line from LARiver to Buffer. The line won't appear until you start dragging.

8) Release the mouse button to connect the input data element to the tool. A menu is displayed to choose which type of connection to make.

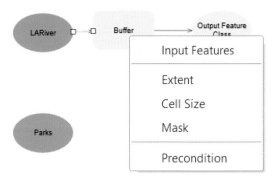

9) On the menu, click Input Features. The tool is now connected to data, but there are other parameters to enter for the Buffer tool so the tool remains gray.

10) Right-click the Buffer tool and click Open (alternatively, you can double-click the tool).

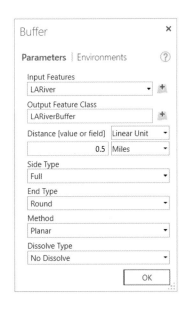

The familiar geoprocessing tool opens, but this time it is floating above ModelBuilder instead of contained in the Geoprocessing pane. The input features are set to LARiver, as specified by the connection arrow, and the output feature class is directed to the default geodatabase. You'll rename this feature class in accordance with your convention.

11) Change the Output Feature Class name to **LARiverBuffer**.

The default output path should be set by the workspace environment settings you made earlier. If not, use the browse button to browse to the ModelOutputs geodatabase.

12) For the linear unit, type **0.5**, and set the adjacent drop-down arrow to **Miles**.

13) Click OK.

In lesson 6, you clicked Run after entering all the parameters to immediately run the tool; in ModelBuilder, clicking OK verifies that the tool is ready to run but doesn't actually run the tool (yet). The tool and output data elements turn yellow and green, respectively, which means they are properly connected and the parameters filled out.

14) On the ModelBuilder tab, in the Model group, click Save.

Buffer the parks

In lesson 6, after buffering the river, you created quarter-mile buffers around existing parks to designate "exclusion zones" when considering where to site a new park.

1) On the ModelBuilder tab, click the Tools button, if the pane isn't open already.

2) Drag another Buffer tool to the model. All the names of the elements (inputs, outputs, and tools) in the model must be unique, so you'll notice that the tool will be named Buffer (2) when you drop it on the model.

3) Position the tool to the right of Parks.

Last time, you set the input features in the Buffer tool by drawing a connector. Now you'll do it another way.

4) Open the new Buffer tool by double-clicking it, or right-click it and click Open.

5) Click the Input Features drop-down arrow and click Parks.

The drop-down list is filled with the input data elements in the model (as opposed to the layers in the map that you've seen in previous lessons). Every input data element (blue ovals) is technically a variable because you can open it and change the feature class it references.

6) Keep the default Output Feature Class name as Parks_Buffer.

7) For Distance, type 0.25. Set the Linear unit drop-down arrow to Miles.

8) Click the Dissolve Type drop-down arrow and click Dissolve all output features into a single feature.

Exercise 7b: Build the model (part 1)   267

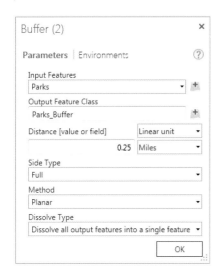

You don't have the visual reason to dissolve the buffers that you had in lesson 6 (you're not going to look at this tool's output, just use it as input for the next step), but it doesn't cause a problem to duplicate the workflow.

9) Compare your dialog box to the figure and click OK.

Both buffer processes should now be colored in your model.

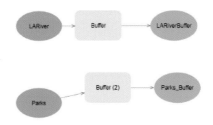

10) Save the model by clicking the Save button on the ModelBuilder tab.

11) Save your project by clicking the Save button at the top of ArcGIS Pro.

Erase park buffers

Now, as in lesson 6, you'll use the Erase tool to subtract the park buffers from the LA River buffer to remove them from consideration for the new park location.

1) If necessary, open the Geoprocessing pane by clicking the Tools button on the ModelBuilder tab.

2) At the top of the Geoprocessing pane, type Erase into the Search box.

3) Drag Erase (Analysis Tools) to the model.

4) Position the tool next to LARiverBuffer.

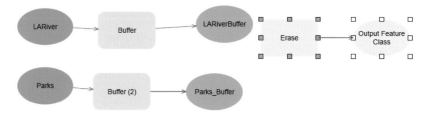

Remember from lesson 6 that the Erase tool requires two input feature classes: (1) the Input Features feature class—the features from which areas will be erased—and (2) the Erase Features feature class—the areas that will be erased (the "cookie cutter").

5) **Connect LARiverBuffer to Erase by drawing a line between them (as you did in steps 6–8 under "Add a tool to ModelBuilder").**

6) **Click Input Features when prompted to define LARiverBuffer as this parameter.**

7) **Connect Parks_Buffer to Erase by drawing a line between them.**

8) **Click Erase Features when prompted.**

The Erase elements will now be colored because Erase doesn't require any other parameters. You will go ahead though and rename the output feature class to give it a more descriptive name and stay consistent with lesson 6.

9) **Double-click Output Feature Class (the output of Erase).**

10) **Change the output name to ProximityZone.**

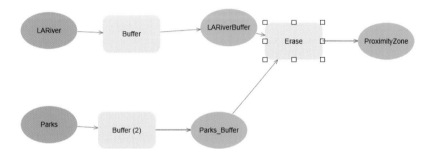

The Erase tool is now ready to go but you can review all the tool parameters just to confirm.

11) **Double-click the Erase tool to open it.**

Note that the tool parameters have been filled in for you by connecting and changing the individual model elements.

12) **Close the dialog box by clicking OK.**

Depending on how you've positioned your elements, your model may be starting to look a little "messy." As you add more elements to the model, it will become increasingly cluttered and difficult to find things. You'll use the Auto Layout tool to clean things up.

Exercise 7b: Build the model (part 1)   269

13) On the ModelBuilder ribbon, in the View group, click the Auto Layout button.

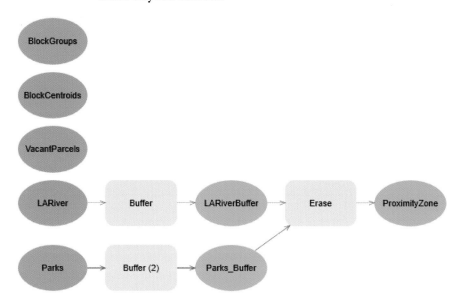

Much better, right? Your resulting layout may look different from the figure, which is fine.

14) Save the model by clicking the Save button on the ModelBuilder tab.

15) Save your project by clicking the Save button at the top of ArcGIS Pro.

Assign block group demographics to the proximity zone

After erasing the park buffers in lesson 6, you used the Identity tool to overlay the remaining area of interest with block groups to assign census attributes. You'll do the same thing now in the model.

1) If necessary, open the Geoprocessing pane by clicking the Tools button on the ModelBuilder tab.

2) At the top of the Geoprocessing pane, type Identity into the Search box.

3) Drag Identity (Analysis Tools) to the model and drop it next to BlockGroups.

4) Double-click Identity to open it.

5) Use the figure to set the parameters.

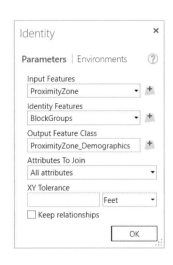

6) **Click OK.**

The Identity tool is now colored in the model.

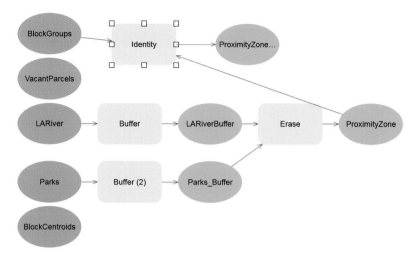

The name of the output of Identity (ProximityZone_Demographics) is too long to fit in the oval so it is being cut off. The text displayed on the elements in the model is just labels and can be changed. Although each label defaults to the tool or data it represents, you can change them to enhance the display and readability of the model.

7) **Right-click the output green oval of the Identity tool and click Rename.**

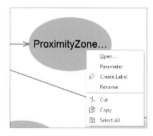

8) **Replace the existing text with Proximity Zone Demographics and press Enter (or click away from the text).**

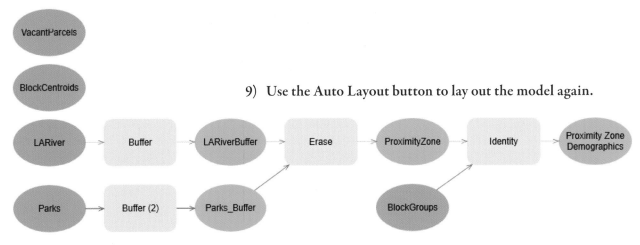

9) Use the Auto Layout button to lay out the model again.

10) Save the model and save your project.

Validate and run the model

This is a good time to test-run the model. Before running a model, you should validate it to make sure that all is well with the data variables and tool parameters.

1) **On the ModelBuilder tab, in the Run group, click the Validate button ✓.**

Validation ensures that a model has access to its source data. If you rename, move, or delete a feature class referenced by an input data variable, that variable and its dependent processes will turn gray (not ready) when you validate the model. Similarly, if you delete a field from a table, and that field is specified by a tool's parameter, the tool and its dependent processes will turn gray when you validate.

▶ If nothing turns gray, your model is ready to run.

The next step is to run the model.

2) **On the ModelBuilder tab, in the Run group, click the Run button ▶.**

As it runs, you should get a visual progress report, with each tool turning red as it is processed. A progress dialog box is also displayed including detailed messages.

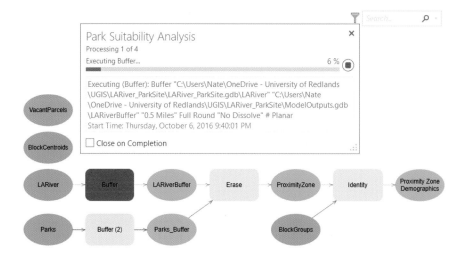

3) **Close the message box when the model is finished running.**

In the Model window, the tools and output data elements have drop shadows. This shadowing tells you which of the tools are complete.

Now that the tools have run and the model is complete, you can view the results and add them to your map.

4) **In the Project pane, browse to the ModelOutputs geodatabase.**

5) **Refresh the geodatabase by right-clicking it and clicking Refresh.**

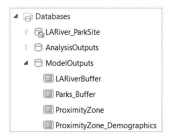

The geodatabase contains four feature classes. They have the names assigned by the tool parameters in ModelBuilder.

6) **Open the Lesson7 map by clicking the tab above the model.** The map is currently empty except for the basemap.

7) **Drag the ProximityZone_Demographics layer to the map.**

8) Open the attribute table of the ProximityZone_ Demographics layer. Scroll across to see its attributes.

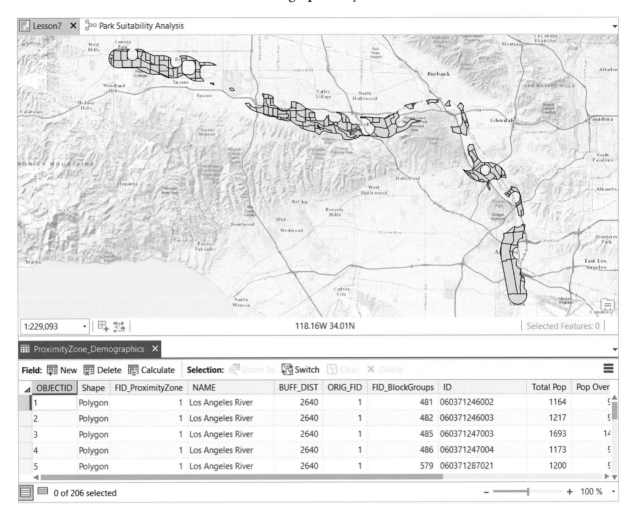

The table has 206 records and includes the demographic attributes from the BlockGroups feature class. It's the same result that you got in exercise 6b. Configuring the Identity tool is the last step in your current model so you can assume the other layers are correct if the Identity output is correct.

9) Remove the ProximityZone_Demographics layer from the Lesson7 map.

10) Switch back to the model by clicking the tab above the map.

11) Save the model and your project.

12) Close the model by closing its tab. Close it even if you're continuing to the next exercise so you can look at how to edit an existing model.

You'll finish the model in exercise 7c. If you want, you can take a break here.

Exercise 7c: Build the model (part 2)

In this exercise, you'll keep following the lesson 6 workflow and finish the analysis model.

In this exercise, you'll see how ModelBuilder handles operations on layers. In lesson 6, you used attribute and spatial queries to narrow down the number of potential park sites. Queries (and, for that matter, any type of feature or record selections) are not made directly on feature classes, but rather on layers. To make feature selections in ModelBuilder, you therefore need a special tool that turns an input feature class into a layer. This tool is appropriately named Make Feature Layer. When you make layers in ModelBuilder, it's usually to make a selection that can be used as an input to further processes. For a specialized use of the Make Feature Layer tool in ArcGIS Pro, see the sidebar "Apportioning attribute values" in lesson 6.

Select features by demographic attributes

In lesson 6, you used an attribute query to select features in the proximity zone according to the demographic values you assigned from block groups using the Identity tool. You'll re-create that query now.

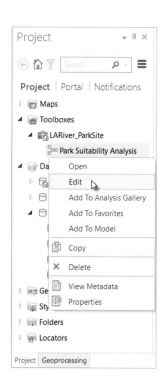

1) If necessary, start ArcGIS Pro and open the LARiver_ParkSite project.

2) In the Project pane, expand the Toolboxes group and find the Park Suitability Analysis model.

3) Right-click the model and click Edit. The model reopens for editing.

4) Open the Tools list and search for Make Feature Layer in the list of tools, and then drag it to the model.

The Make Feature Layer tool is similar to Select By Attributes, but instead of selecting and highlighting the features that satisfy an expression, it creates a temporary layer that filters the features. Both tools allow you to define a query to limit the features that are used in subsequent tools. In this example, you'll use the tool to define the demographic criteria to be used in the next steps.

5) Double-click the tool to open it.

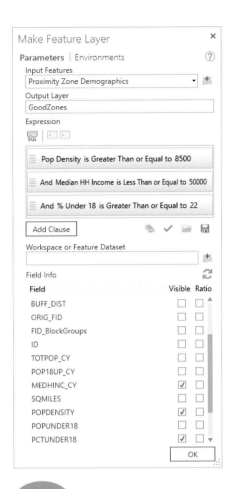

6) Set Input Features to Proximity Zone Demographics.

7) Change the Output Layer name to Good Zones.

8) Add the clauses to select features in which all the following conditions are true (as you did in lesson 6):
   - Pop Density is greater than or equal to 8500
   - And Median HH Income is less than or equal to 50000
   - And % Under 18 is greater than or equal to 22

Make Feature Layer also gives you the opportunity to hide unnecessary fields so you'll go ahead and remove the extra fields that were added from the buffer steps in exercise 7b.

9) Hide all the fields except for the following by clearing the Visible check box:
   - OBJECTID
   - Shape
   - MEDHINC_CY
   - POPDENSITY
   - PCTUNDER18

These fields are the only ones you need for your analysis, so by hiding the others at this point, they'll be excluded from future analysis in the model.

10) Compare to the figure and click OK.

11) Use the Auto Layout button to lay out the model again.

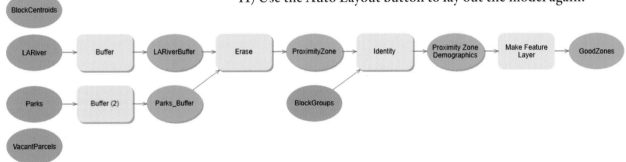

The ModelBuilder tab has several navigation tools in the View and Mode groups for panning and zooming in and out of your model.

You can also use the scroll wheel as well as the horizontal and vertical scroll bars to navigate the model.

12) Save the model and your project.

Select parcels within good zones

The attribute query for desirable demographics significantly narrows down the "good zones" in which you can look for vacant parcels. You now have the ModelBuilder equivalent of a layer with a definition expression set, bringing you to the point in the analysis where you were at the end of exercise 6b. The next thing to do is make a spatial query for vacant parcels that lie within the good zones.

1) Search for Make Feature Layer and drag it to the model next to the VacantParcels element.

2) Drag to draw a line from VacantParcels to Make Feature Layer (2) and set it as the Input Features parameter.

3) Open the Make Feature Layer (2) tool.

In lesson 6, you did a Select By Location query for vacant parcels in good zones, followed by an attribute query for parcels a quarter acre or larger. You're going to now switch the order of these two query operations by adding the attribute expression to the Make Feature Layer tool, but it doesn't affect the logic. It doesn't matter whether you first select the vacant parcels in good zones and then select the quarter-acre ones (lesson 6) or first select all the quarter-acre vacant parcels and then select the ones in good zones (as you'll do here).

4) Change the Output Layer name to Suitable Size Vacant Parcels.

5) Add a clause for:

Acres is greater than or equal to 0.25.

6) Compare your tool to the figure and click OK.

7) Search for the Select Layer By Location tool and drag it to the model next to Suitable Size Vacant Parcels.

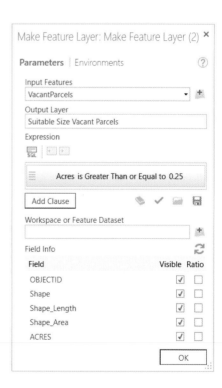

8) Open the Select Layer By Location tool and use the figure to complete the parameters to select the vacant parcels that are completely within the good zones. Click OK.

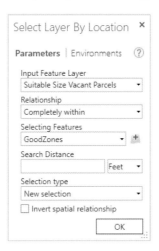

9) Click the Auto Layout button.

10) Rename the last output data element Good Zone Parcel Layer.

Note that the last output that actually creates a feature class in the geodatabase is Proximity Zone Demographics. You can tell by the shadowing of the last green oval. The subsequent outputs consist of two layers created by the Make Feature Layer tools and selection on those layers.

11) Hover over the Proximity Zone Demographics element.

The ToolTip shows the path to the dataset on disk.

12) Hover over each of the downstream outputs in turn.

These ToolTips show only the name of the layer, because that's all there is to see—no datasets have been created on disk. There are four output elements, but only two layers: Good Zones, created by Make Feature Layer, and Suitable Size Vacant Parcels layer, created by Make Feature Layer (2). Good Zone Parcel Layer is a spatial selection of Vacant Parcels. You renamed the output elements of the selection tools for clarity (hopefully), but you didn't have to.

Copy features

Just as in lesson 6, you'll use the Copy Features tool to save the selected Good Zone Parcels as a new feature class in the geodatabase.

1) Search for Copy Features in the Tools list and drag the tool to the model.

2) Drag to draw a line from Good Zone Parcel Layer to the Copy Features tool and set it as the Input Features parameter.

3) Open the tool and change the default Output Feature Class name to GoodZoneParcels.

4) Compare your dialog box to the figure and click OK.

5) Click the Auto Layout button.

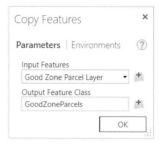

6) Save the model and your project.

This feature class contains your recommended sites, but you'll add a little more information about the sites to be included in the final output.

Determine the distance of each site to the river

You know that each site is within a half mile of the river, but just as in lesson 6, you'll use the Near tool to get a precise distance to help the decision-makers on the city council weigh their options.

1) Search for the Near tool and drag it to the model close to GoodZoneParcels.

2) Open the Near tool and use the drop-down menu to set Input Features to GoodZoneParcels.

3) Set the Near Features parameter to LARiver.

As a reminder, there is no output feature class when using the Near tool because it simply adds fields to the input feature class.

4) Change the Method parameter to Geodesic.

5) Compare the dialog box to the figure and click OK.

Exercise 7c: Build the model (part 2) 279

6) Click the Auto Layout button.

7) Rename the last output element Good Zone Parcels with River Distance.

Determine park access population

Park access is defined as the population within a quarter mile of the site. You made this calculation using the Spatial Join tool with a "within a distance" operator and a "merge rule" in lesson 6.

1) In the list of tools, search for Spatial Join, and drag the tool to the model next to Good Zone Parcels with River Distance.

2) Open the tool and set Target Features to Good Zone Parcels with River Distance.

3) Set Join Features to BlockCentroids.

4) Change the Output Feature Class name to Sites_Access. This will be your final output of the model.

You'll delete unnecessary fields from the output attribute table. Make sure to delete carefully so that your mouse clicks don't get ahead of you.

5) In the list of output fields, click the first field in the list (it should be Shape_Length) to select it.

6) Click the Remove button ✗ on the right of the field name to remove the field from the list (and therefore from the output of the tool).

7) Go on to remove all the fields except ACRES, NEAR_DIST, and POP2010.

▶ If you accidentally delete one of these fields, you can click the Reset button ↻ in the upper-right corner of the table.

8) In the list of output fields, click NEAR_DIST and then the Properties tab (under the Reset button).

9) Change both Field Name and Alias to RiverDistance.

10) Click POP2010, and on the Properties tab, change Field Name and Alias to AccessPopulation.

11) With AccessPopulation still selected in the Output Fields list, click the Source tab (left of the Properties tab), and then click Sum in the Merge Rule drop-down list.

Match Option defines the spatial relationship to be evaluated. You must evaluate the block centroids within a quarter mile of the sites.

12) Change Match Option to Within a distance.

13) Set the search radius to 0.25 miles.

14) Compare your tool to the figure and click OK.

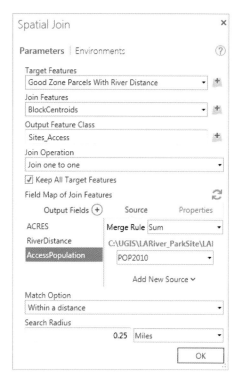

This setup essentially means that the POP2010 values of all the BlockCentroids points within a quarter mile of each site will be summed and included in the output in the POP2010 field.

15) Rename the output element of the Spatial Join tool from Sites_Access to Sites With Access Population.

16) Click the Auto Layout button.

17) Save the model and your project.

Attach the demographic attributes

You're nearly done with the model. Now you'll overlay the sites with the block groups to get the demographics attributes you need. You already know that these sites meet the demographic criteria, but adding the individual numbers for each site will help decision-makers evaluate the sites at a glance.

1) In the list of tools, search for Identity, and drag the tool to the model next to Sites With Access Population.

2) Drag to draw a line from Sites With Access Population to connect it to the tool as the Input Features parameter.

Exercise 7c: Build the model (part 2)   281

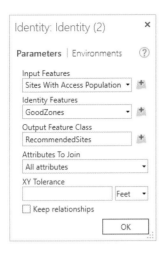

3) Drag to draw a line from the GoodZones layer to connect it to the tool as the Identity Features parameter. The GoodZones layer is a great input because it already has all the unnecessary attributes hidden for you. This will save you some cleanup steps.

4) Open the new Identity tool and change the Output Feature Class name to RecommendedSites.

5) Compare your tool to the figure and click OK.

6) Rename the RecommendedSites element at the end of the model Output Recommended Sites.

7) Click the Auto Layout button.

Run the model

You're ready to run the model.

1) On the ModelBuilder tab, click the Run button.

2) When the model is finished running, close the message box. All the elements should now be displayed with gray drop shadows.

▶ If the model did not finish successfully for any reason, look for the first tool without a drop shadow—this is your problem area to fix.

View the results

1) Open the Lesson7 map by clicking the tab above the model (or from the Project pane if it's not open as a tab).

2) In the Project pane, expand the ModelOutputs geodatabase and drag RecommendedSites to the map. Just as in lesson 6, the sites will be difficult to see at this scale.

▶ If you don't see all the outputs, refresh the geodatabase by right-clicking it and clicking Refresh.

3) Open the RecommendedSites attribute table.

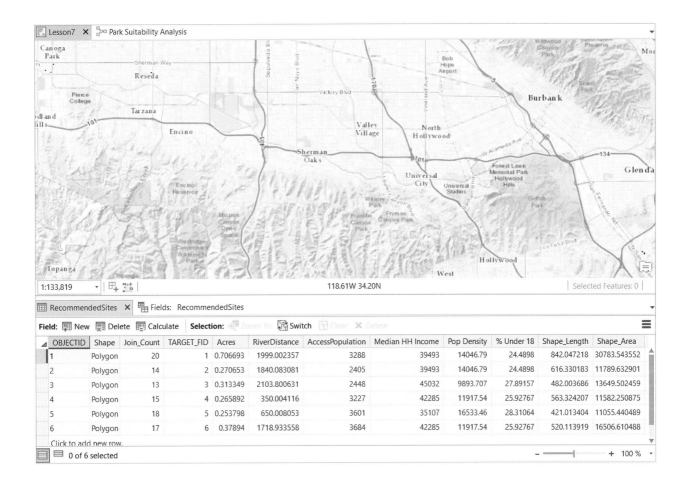

You should have the same six sites with the same Acres, river distance, and access populations that you got in lesson 6. You also ended up with two unnecessary fields as a result of the spatial join, but you'll leave those for now.

Exercise 7d: Run the model as a tool

In this exercise, you'll set up the model as a geoprocessing tool. This is an efficient way to run a model, and it's a good way to share the model with people who may want to use it but are not so interested in the details of its construction. When you run a model as a tool, it has a dialog box interface like any other geoprocessing tool: the user sets parameters and clicks Run. Behind the scenes, of course, the model is a long chain of tools, each of which has its own required and optional parameters. If all these parameters had to be set by the user in the model's dialog box, it would be a long process.

Fortunately, that's not how it works. When you set up the model to run as a tool, you specify which parameters from which tools inside the model will be displayed in the model dialog box. Only these parameters are filled out by the user (you can supply default values). All others are preset by you, and the user running the tool never sees them. In other words, you decide how flexible to make the tool.

The ability to test sensitivity to changes in variables and to explore alternative outcomes are key reasons for building models in the first place. Accordingly, when you set up the model to run as a tool, you should think about which factors the user is most likely to want to experiment with.

Open the model

In previous exercises, you've edited the model by right-clicking it. To run the model as a geoprocessing tool, you open it.

1) If necessary, start ArcGIS Pro and open the LARiver_ParkSite project.

2) In the Project pane, expand Toolboxes if necessary and double-click Park Suitability Analysis.

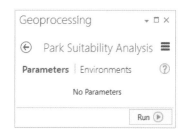

The model opens as a geoprocessing tool similar to all the other tools you've been using in this book. It's just an empty gray dialog box with the message, "No Parameters." That's because you haven't configured which parameters you want to show here yet. The Run button is displayed, however, and you can run it directly from here, and it will use all the model parameters you configured in the previous exercises.

3) Close the Geoprocessing pane.

4) In the Project pane, right-click the model and click Edit.

5) In the Model window, zoom in on the left side of the model and set your view scale to a comfortable level.

Make a tool parameter for buffer size

Users of the model tool will probably want to experiment with the buffer sizes for maximum distance from the river (0.5 miles) and minimum distance from existing parks (0.25 miles). To display these distances as parameters when you run the model as a tool, you must first define them as variables. This configuration allows their values to be changed any time the model is run. Then you flag the variables as "parameters" (parameters you want to include in the model's tool dialog box).

1) In the Model window, right-click the Buffer tool connected to the LARiver input, and click Create Variable > From Parameter > Distance [value or field].

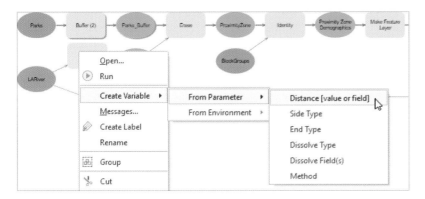

This setting says that you want the Distance parameter in the Buffer tool to be a variable in ModelBuilder. A small light-blue ellipse, called a *value variable*, is added to the Model window.

2) Move the variable a bit so it's not overlapping any other elements.

3) Double-click the variable to open its dialog box.

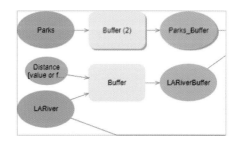

Within the Model window, the Distance parameter is now detached from the Buffer tool and can only be set here. You'll accept the current value of 0.5 miles that you set earlier, in exercise 7b, as the default for the model tool.

4) Close the Distance dialog box by clicking OK or clicking the X in the upper-right corner.

5) Right-click the value variable and click Parameter.

The letter *P* (for parameter) appears above the variable.

The variable name (in the oval) is important for tool parameters. This name will appear as the label for the parameter on the tool. You don't really want the label to ask for "Distance [value or field]." A more descriptive name would be "Distance to LA River."

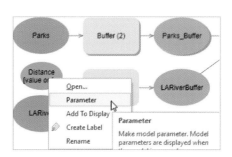

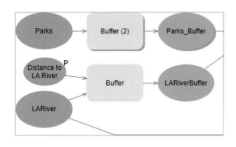

6) Rename the Distance [value or field] variable Distance to LA River.

You'll make a parameter for the Parks buffer distance in the same way.

7) Right-click the Buffer (2) tool connected to the Parks input, and click Create Variable > From Parameter > Distance [value or field].

8) Move it so it doesn't overlap other elements.

9) Right-click the variable and click Parameter.

10) Rename the variable Distance from Parks.

11) Save the model and your project.

12) In the Project pane, in the project toolbox, double-click the Park Suitability Analysis model again and look at the resulting tool dialog box.

Note that the two buffer distance parameters have been added to the dialog box, along with the variable names and default values. These distances can now be changed and the model run, but you'll wait and create a few more parameters first.

13) Close the Geoprocessing pane.

Make parameters for demographics and park size

You'll make tool parameters from the query expressions for demographics and park size so you can experiment with these parameters, too.

1) Right-click the Make Feature Layer tool that comes from Proximity Zone Demographics, and click Create Variable > From Parameter > Expression.

A value variable with the name Expression is added to the Model window.

2) Move the new variable to a good spot.

3) Rename the variable Demographic Criteria.

4) Flag this variable as a parameter

5) Right-click the Make Feature Layer (2) tool connected to the Vacant Parcels input, and click Create Variable > From Parameter > Expression.

6) Move the new variable to a good spot.

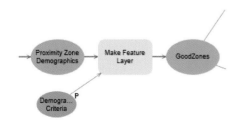

7) Hover over the element to see its expression (ACRES >= .25). This tip is displayed as a SQL statement (which is how the clauses you created are ultimately evaluated by ArcGIS Pro).

8) Rename the variable Park Size in Acres.

9) Flag the variable as a model parameter.

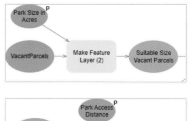

10) Scroll to the right, if necessary, to right-click the Spatial Join tool connected to the BlockCentroids input, and click Create Variable > From Parameter > Search Radius.

11) Move the new variable to a good spot.

12) Rename the variable Park Access Distance.

13) Flag the variable as a parameter.

14) Use the Auto Layout button to clean up the model layout.

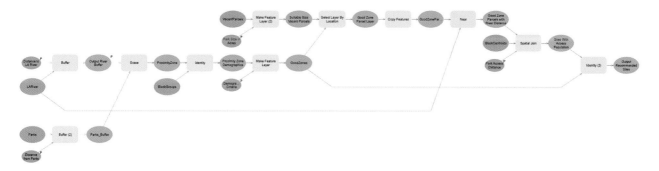

15) Click the Validate Entire Model button ✓.

16) Save the model and your project.

At this point, you are done creating the model and can run it as a tool. When models are configured properly with parameters, they run just like any of the other tools in the toolbox (such as Buffer, Clip, Spatial Join, and so on). The user of the tool simply provides a few parameters to run it and never has to open up the model or even really understand how it works. You've now configured the model with enough parameters to run it quickly and easily as a tool. You will also add a few more parameters to give users even more control over how to run the analysis.

Run the tool

Now you'll rerun the model as a tool.

1) Switch to the Lesson7 map tab (or open the Lesson7 map, if necessary) so that it is displayed.

2) In the Project pane, double-click the Park Suitability Analysis tool in your project toolbox.

The tool displays the parameters you've specified.

3) Click the Run button to run the tool with the default parameters that you set in the model.

A progress bar and status messages are displayed at the bottom of the Geoprocessing pane.

4) When the tool is finished running, expand and refresh the ModelOutputs geodatabase.

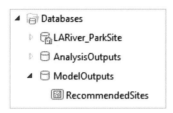

Note that the only feature class in the geodatabase now is RecommendedSites. This is because ArcGIS Pro assumes that the only outputs to be saved are either

- the very last output in a chain or
- outputs flagged as a parameter.

All other datasets created during the execution of the tool are considered "intermediate" and are deleted after the tool runs successfully. Therefore, any output datasets created by your model that might be useful should be flagged as a parameter.

In your model's current state, you probably want to keep the LARiverBuffer feature class, so you'll make that a parameter in a later step in this exercise.

5) Open the Park Suitability Analysis tool and double the acceptable distance to the river by changing the Distance to LA River parameter to 1.

▶ To open the tool, as opposed to the model of the same name, double-click the tool in the Project pane.

6) Run the tool again.

7) When the tool is finished running, remove the RecommendedSites layer from the Lesson7 map and add it again from the ModelOutputs geodatabase.

8) Open the RecommendedSites attribute table.

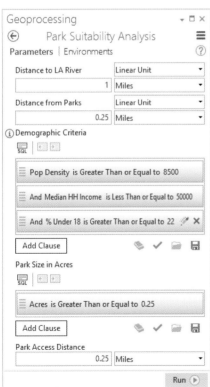

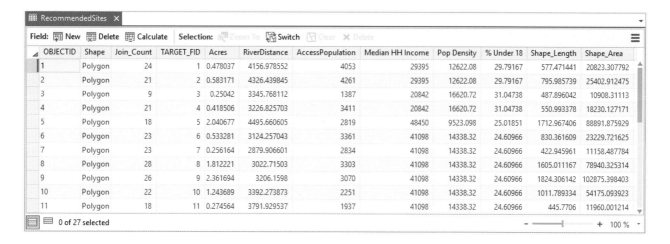

By doubling the distance from the river that you consider acceptable for a new park, you now have 27 recommended parks instead of your original six. Now imagine making this same minor modification to the step-by-step analysis you did in lesson 6—this relatively minor change would have taken a long time to do and would leave several opportunities for other mistakes to creep up in the process.

Make tool parameters for the outputs

Flagging the outputs of the model as tool parameters will allow you to keep the outputs after the process tool runs and add the results as layers to your map. More importantly, it will give you control over where to save your outputs and whether they should overwrite each other. As you experiment with different values in successive runs of the model, you may want to save your results with different names so you can compare them.

1) Switch to the model view of Park Suitability Analysis (go to the tab if it is still open, or right-click the model in the Project pane and click Edit).

2) In the Model window, right-click the LARiverBuffer output data element and click Parameter.

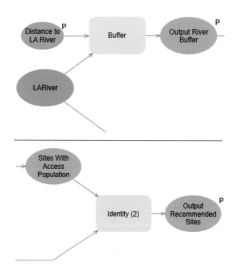

You don't have to create a variable first as you did with the other parameters because data elements are variables by definition.

Because you're now using this element as a parameter, you can give it a cleaner label so it looks more polished in the tool dialog box.

3) Rename LARiverBuffer as **Output River Buffer**.

4) Scroll to the far right of the model, and set Output Recommended Sites as a parameter.

Running the tool versus running the model

Running a model (from the ModelBuilder ribbon) runs all the processes using the default values. Running the tool (from the Project pane) runs things using the values you enter into the parameter boxes. Most models are ultimately meant to be run as tools by users, who will probably never actually open the model to see how it works.

5) Validate the model, and then save the model and your project.

6) Close the model view by closing the Park Suitability Analysis tab.

Run the tool using different values

Now you'll run the model using some different parameter values.

1) Open the Lesson7 map, if it's not open already.

2) In the Project pane, in your project toolbox, double-click the Park Suitability Analysis tool. The tool appears with all the parameters you set in the model.

3) Run the tool with all the default parameter values. Note that the output data that you configured as parameters is now being added to the map as layers.

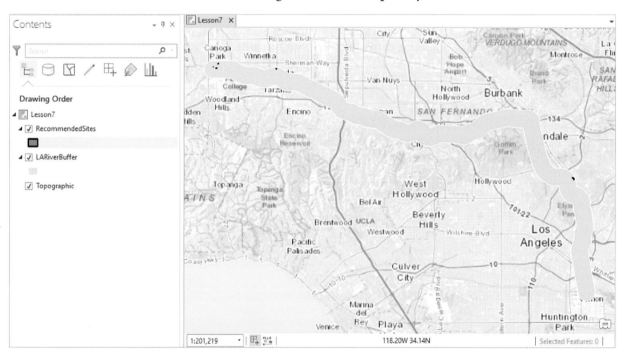

Now you'll run the tool again using different values and save the resulting feature classes with different names.

4) Change the Distance to LA River value to 1 (miles).

5) Edit the Park Size in Acres clause to Acres is Greater Than or Equal to 0.5.

290 Lesson 7: Automate the analysis

6) Append the text _OneMile (for example, LARiverBuffer_ OneMile) to the end of the two output feature class names.

7) Compare your tool to the figure and click Run.

8) Close the Geoprocessing pane when the tool is finished running.

The two new layers are added to the Lesson7 map. You should now see the one mile and larger park results (_OneMile) displayed on top of the original results you created.

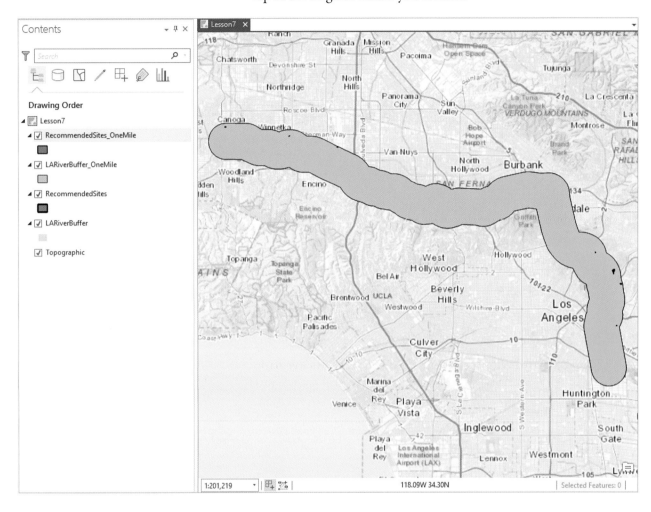

9) Turn the _OneMile results on and off to compare with the original results. There are 11 sites that match the new criteria (potential park sites must be larger but can be farther away from the river).

▶ Open the attribute table of RecommendedSites_OneMile to see the 11 new sites.

10) In the Project pane, expand ModelOutputs.

292   Lesson 7: Automate the analysis

The geodatabase should contain four feature classes: two from this run of the model and two from the previous run.

11) **Experiment with your tool by adjusting the other parameter values.**

12) **Save and close your project.**

You've now seen two ways to carry out an analysis project: as a series of stand-alone geoprocessing operations and as a model that can also be used as a tool. Models offer a lot of advantages, but it's an open question as to how fully developed the analysis should be before you turn it into a model.

We recommend planning the analysis in a traditional way: sketching the workflow on paper or a whiteboard, experimenting with geoprocessing tools, and discussing the results with colleagues. Wait until you have a firm plan before you take the analysis into ModelBuilder. (And remember that even your "firm plan" will probably change.) If you start with ModelBuilder as your whiteboard, you may find that the model gets too messy in the early planning stages and needs constant corrections. Of course, the more you know about analysis, geoprocessing tools, and ModelBuilder, the sooner you can get your model in good working order. Once that's done, it's relatively easy to add further complexity.

Now that you've run the analysis and identified some suitable park sites, you'll need to communicate your results with stakeholders and decision-makers through map products. In lesson 8, you'll create a printable map layout with essential cartographic elements such as a title, scale bar, north arrow, and legend.

Lesson

8 Present your analysis results

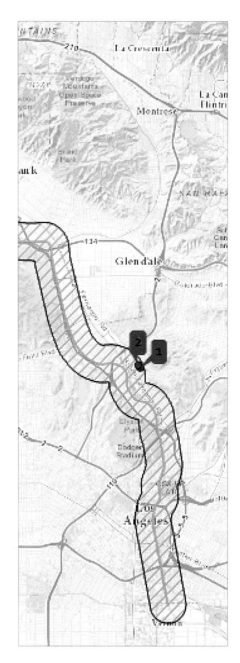

COMMUNICATING YOUR FINDINGS

to a diverse audience of readers, with levels of sophistication ranging from public citizens to council members to the city's urban planners who must ultimately design an actual park, is no easy task.

Your analysis has yielded a short list of candidate sites that you want to present to the city council and other interested parties. You'll make your presentation in the form of a map that places the results in a meaningful geographic context, addresses the project guidelines, and follows good cartographic design principles.

The deliverable item is an 8½ x 11-inch map that can be printed or viewed on-screen.

But wait, you may be thinking, what's wrong with simply adding the layer of recommended sites to an online basemap, as you already did in lesson 6? There's nothing wrong with it, and basemaps will be part of your map design, but your map design must accomplish a lot of things. You'll be adding inset maps, customizing the positions of labels and other text, and incorporating other standard cartographic elements such as a scale bar and north arrow.

Your analysis results aren't too complex. You basically want to show just the study area, represented by the LA River buffer zone, and the six proposed park boundaries. You want to set these analytical layers against a backdrop of topography, place-names, and major roads. Perhaps the biggest design challenge you face is that these six locations are small parcels of land strung out over a 20-mile stretch of river corridor: How do you map the area of interest without rendering the parks too tiny to see at that scale? The solution lies in the use of inset maps—smaller maps placed within the main map, each one portraying a closeup view of a particular site.

Cartography involves innumerable choices about color, font, line width, the placement and juxtaposition of elements, and so on. These are often somewhat subjective decisions, with many good (and bad!) possibilities. This lesson takes you through the creation

Lesson Eight road map

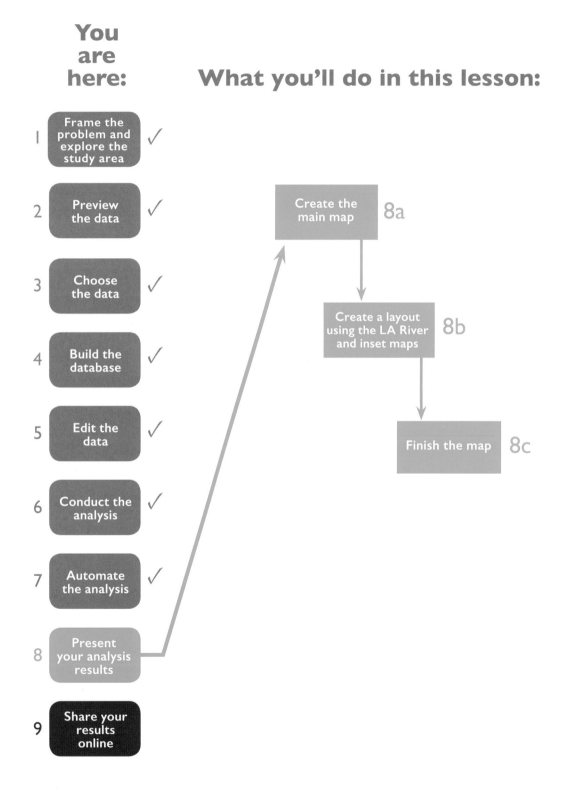

of an exact design (shown in the figure at 60 percent actual size), but along the way you'll have the option of altering things to suit your own taste. We'll try to explain why we made the choices we did, but you don't have to follow the instructions with total fidelity. Experimenting with some of the settings will give you a deeper understanding of how the software works and a chance to make a map that suits your own aesthetic tastes.

A full presentation of this type in the real world might include a variety of media products, such as large-format wall maps, smaller notebook-size maps, printed reports, digital slide shows, PDF downloads, and websites. We can't do all these things without writing another book, so we're going to focus on a single map layout that can be printed on a standard letter-size piece of paper. In lesson 9, you'll go a step further and take your results to the web, but for now, your working medium will be a sheet of letter-size paper.

An 8½ by 11-inch map is small, but it would be worse if you were trying to pack a lot more features or a lot of detailed data into this format. We're using this size in recognition of the fact that many people don't have access to larger-format printing devices. And if you can design well in small format, you'll feel that much more comfortable and creative when you begin working in larger formats.

A good way to begin a map layout—just like an analysis—is with a sketch of your basic plan. The sketch for your map is shown in the figure.

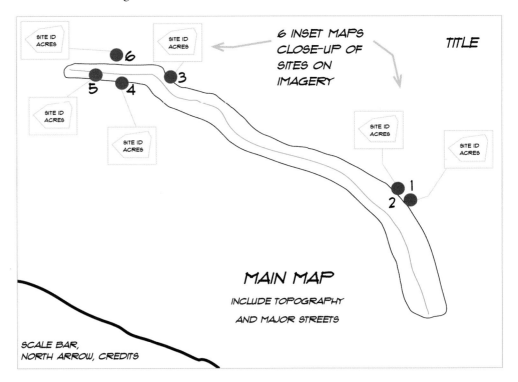

Exercise 8a: Create the main map

Beginning with this exercise, you'll create a new map that will be used for the map layout.

Create a new map for layout

1) Start ArcGIS Pro and open the LARiver_ParkSite project.
2) Close any maps or tables that may be open.
3) Make a copy of Lesson6e and rename it Lesson8.
4) Open the Lesson8 map.

5) Remove the Bright Map Notes layer.

6) Use a blue symbol for the LARiver layer (if one is not already set).

7) Zoom to the LARiverBuffer layer.

8) Change the basemap to Topographic.

9) Change the map projection of Lesson8 to WGS 1984 Web Mercator Auxiliary Sphere. This option should be under Map Properties > Coordinate System > Layers.

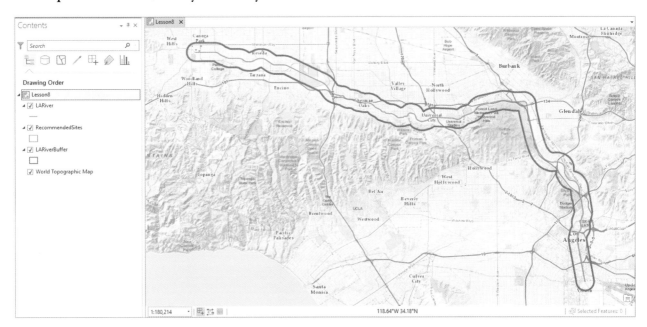

The World Topographic Map basemap will make an unobtrusive background, with topographic relief that will make the map less starkly flat. All the basemaps in Pro are based on WGS 1984 Web Mercator Auxiliary Sphere and therefore will display the least amount of distortion of features and text for your map layout.

Symbolize the LA River buffer

In lesson 6, you created the LARiverBuffer feature class. You'll now symbolize it using a hatched fill symbol for use in your layout.

1) Open the Symbology pane for the LARiverBuffer.

2) In the Symbology pane, click on the symbol to open symbol formatting and, if necessary, switch to the Properties tab and then the Layers view .

Exercise 8a: Create the main map   299

3) Keep the line as a solid stroke, but change the color to black and the width to 1 point.

4) Select the buffer polygon, and change the drop-down list next to Solid fill to Hatched fill.

5) Change the color to a green.

6) Change Separation to 6 pt (this option is located in the Pattern drop-down list). Separation is the distance between each of the hatch lines.

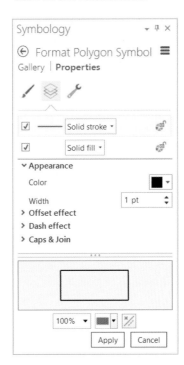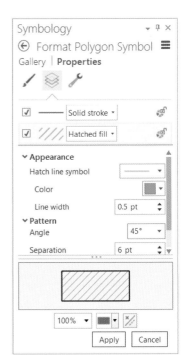

7) Click Apply.

Why make the buffer green? One reason is that you have many layers to symbolize, and green works well in combination with the other colors you're going to use. On a psychological level, by using green you're also suggesting that this area is verdant, riparian habitat—whereas, in reality, much of it is either densely developed residential/commercial land or industrial blight. Are you misleading your audience, or just being optimistic? Whichever way you think of it, you must be aware as a cartographer that symbols may have hidden meanings as well as overt ones.

Add and symbolize the LA city limits

You only considered locations within Los Angeles itself, so it will be helpful to show the city limits on the map.

1) In the Project pane, browse to your MapsAndMore folder, drag the Los Angeles.lyrx file to Contents, and drop it right above the Basemap layer (a horizontal gray bar marks the spot).

When you drop a layer on the map, it takes its natural place in the top-to-bottom drawing order (point, line, polygon, and raster). When you drop it in the Contents pane, it goes exactly where you put it.

2) Change the symbol for the new Los Angeles layer to use the Yellow (Bright) symbol from the Gallery view.

3) On the Appearance ribbon, change Layer Transparency to 95%.

The result is a subdued yellow mask on top of the basemap. White space shows through areas that aren't part of Los Angeles proper.

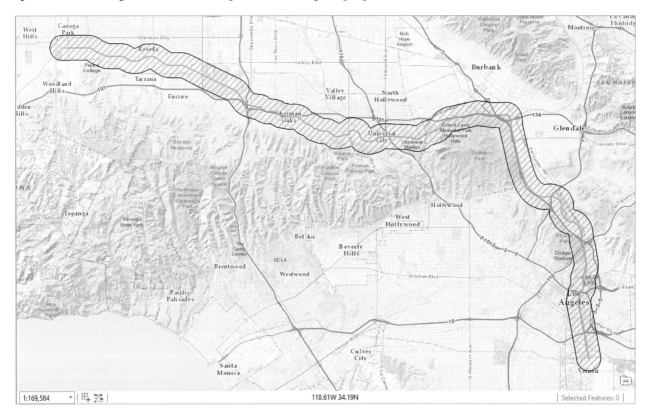

Exercise 8a: Create the main map 301

Set the map extent

At the moment, the map is zoomed to the extent of the LARiverBuffer layer. Although this is your area of interest, you must make a little more room around the edges for inset maps.

1) **Below the map, set the map scale to 1:160,000.**

At this scale, one inch on the map equals about 2½ miles on the ground. With your ArcGIS Pro window maximized, the map extent should be close to that of the figure, although the result depends on such things as the size and screen resolution of your monitor. For your purposes, the extent is more important than the scale.

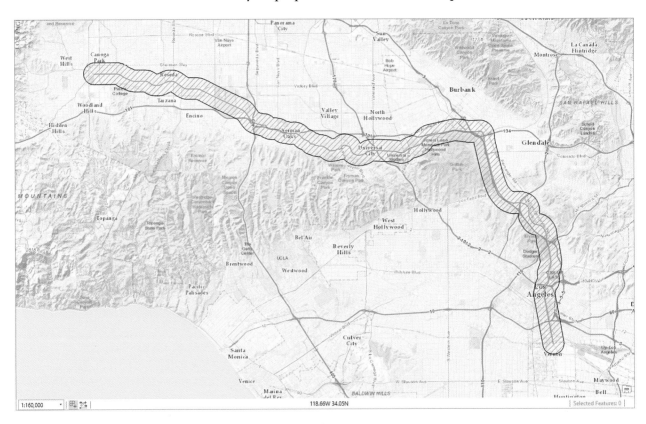

2) **On the Map tab, in the Navigate group, create a bookmark named LA River Layout Extent.**

Symbolize recommended sites

The six recommended sites are the most important features on your map, but they're so small and close together that you must do some creative symbology to make them show up. In our sketch at the beginning of the lesson, we used red circles to represent the site locations and balloon labels to designate Site ID and acres. You'll create these effects here by symbolizing the RecommendedSites layer (a polygon layer) as a point marker and create callout text symbols. Because you can easily enlarge the symbols to a size that will draw the map reader's attention, you'll use them as substitutes for the small-site polygons. It will still be important to show the actual shape and size of the sites, and to do that, you'll create inset maps in exercise 8b.

1) Turn on the RecommendedSites layer, if necessary, and open its attribute table.
2) Confirm that it shows the field aliases and formatting from lesson 6 and close the table.

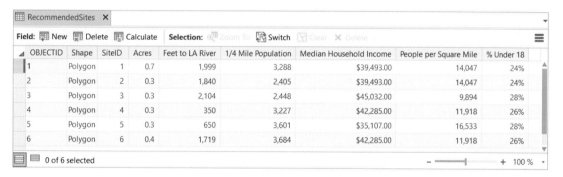

3) Open Symbology for RecommendedSites.
4) In the Symbology drop-down menu, confirm that Symbology is set to Single Symbol.
5) Select the symbol to open Format Polygon Symbol.
6) Click Properties and then click the Structure button ⚒.
7) Under Layers, click Add symbol layer and then Marker layer.

8) Delete the line and polygon layers by clicking the Delete Layer button ✕ next to each layer, leaving only the Marker layer.

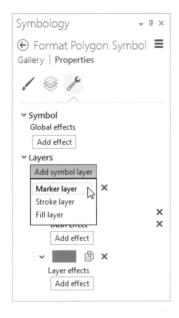 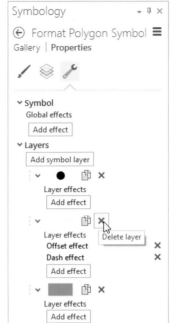 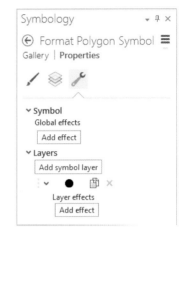

9) Switch to the Layers tab.

10) Under Appearance > Form, click Style and then Circle 3. Click OK.

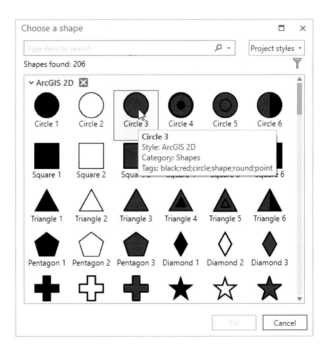

11) Change the size to 6 point.

12) Expand Marker Placement and change Placement to At Center.

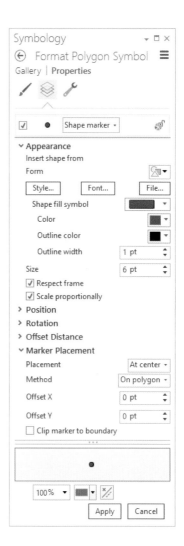

13) Compare to the figure and click Apply.

Each of the six sites is symbolized with red circles at the center of each polygon.

On the map, the symbols stand out prominently. Cartographers often use bright colors (called "spot colors") for the features they most want to call attention to on the map.

Label the sites

You'll use the SiteID field to label the sites.

1) Highlight RecommendedSites, if necessary, and click the Labeling tab. On the far left of the ribbon, in the Layer group, click the Label button ✏️ to enable it.

2) Just to the right of the Label button, in the Label Class group, click the Field drop-down arrow and click SiteID.

3) In the Text Symbol group, expand the gallery and scroll down to Layout. Click Callout (Sans Serif).

4) Click the Text Symbol button ⌐ in the lower-right corner of the Text Symbol group to open the Label Class pane.

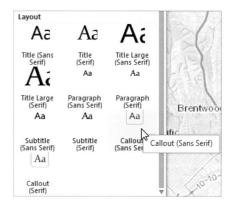

Exercise 8a: Create the main map   305

5) Confirm that the Symbol tab is selected at the top of the pane (between Class and Position) and that the General tab is selected (below Class).

6) Expand Callout and change the color to Poinsettia Red to match the Marker layer color. Click Apply.

The labels are now drawing with the correct style but are overlapping in areas in which you have groups of sites, such as sites 4, 5, and 6. You'll apply an offset to draw leader lines to the sites when needed.

7) Click the Position tab at the top of the pane and expand the Placement group.

8) Set the Preferred offset to 10.0. When you are finished, close the Label Class pane.

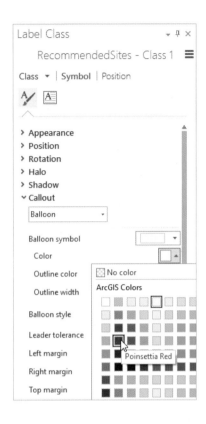

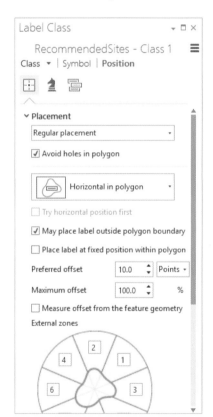

9) Zoom to the LA River Layout Extent bookmark. Your Lesson8 map should look similar to the figure. If all the labels aren't displayed, try zooming in.

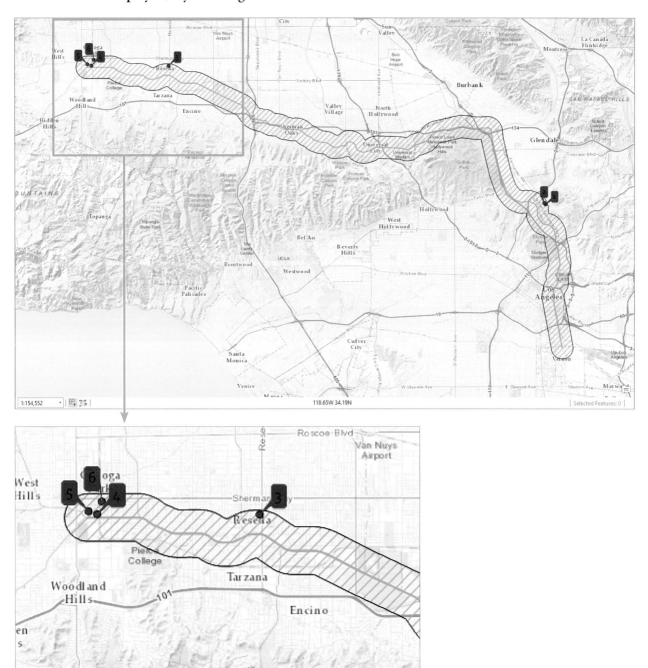

10) Save your project.

Exercise 8a: Create the main map   307

11) **If you're continuing to the next exercise, leave ArcGIS Pro open. Otherwise, close ArcGIS Pro.**

In this lesson, you've symbolized the layers that make up the Lesson8 map. In the next exercise, you'll create a new layout using both an LA River map and inset maps to show close-ups of the sites.

Exercise 8b: Create a layout using the LA River and inset maps

You'll now be working in an ArcGIS Pro layout. Viewing and working with data is one thing; composing a map for the printed page is something else. In composing a map, you must take into account not only your design considerations but also output constraints such as paper size, page orientation, and print margins. Map composition is done in a layout—the graphical, page layout environment of ArcGIS Pro.

Create a new layout frame

1) On the Insert tab, in the Project group, click New Layout.

2) Under the ANSI – Landscape section, click Letter. In the Contents pane, rename the new layout Lesson8. Also note that the new layout has been added to the Layouts folder in the Project pane.

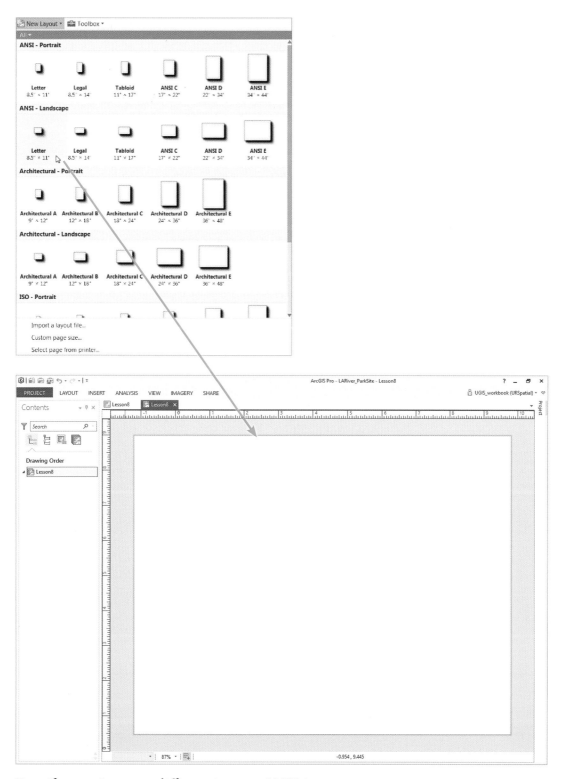

Even if your printer uses different sizes, use ANSI A anyway to complete the exercise.

Exercise 8b: Create a layout using the LA River and inset maps

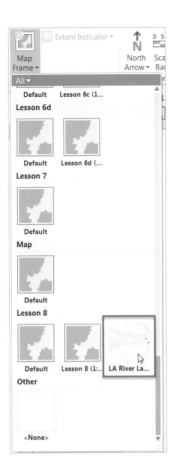

3) **Insert a map frame using the LA River Layout Extent bookmark.**

The idea of a map frame takes on new significance in a layout. In a map (where you've worked until now), a map frame is an abstract "container" of layers. Apart from one or two special properties, such as its coordinate system setting that controls on-the-fly projection, you generally don't pay much attention to it. In a layout, the map frame is visual: it's a rectangular window placed on a virtual sheet of paper and serves as a key element in the map composition.

In the template, the default margins around the map frame are set to one inch on all sides. Most of this space would be wasted in your layout because you're going to place all the map content—data and marginalia both—directly in the map frame. You'll enlarge the map frame, leaving just a minimum amount of white space around the edge of the page.

4) **Under Contents, right-click Lesson8 Map Frame and click Properties.**

5) **In the Format Map Frame pane, click Placement and change the width to 10.5 and height to 8.**

6) **Change X to 0.25 and Y to 0.25.**

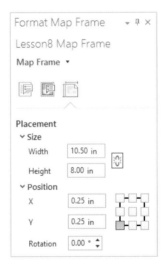

7) **Close the Format Map Frame pane.**

8) **Click the Layout tab.**

Now is a good time to make the distinction between the navigation tools on the Layout tab for the map frame and the map layout itself.

310   Lesson 8: Present your analysis results

The icons for navigation tools on the Layout tab, in the Navigate group, show an 8½ x 11-inch "page" in the background of the layout. These are the tools used to navigate around the actual map layout.

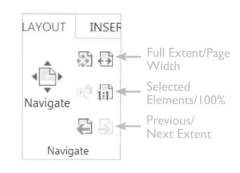

The Layout navigation tools magnify, reduce, and otherwise adjust your view of the map layout. They do not change the scale or extent of the map within the map frame.

- Try zooming in and out of the page using the scroll wheel.
- Pan using the left mouse button. Press the Shift key to enable a Zoom In tool.
- Use the Full Extent button to return to the entire map layout.

9) **On the Layout tab, in the Map group, click the Activate button.**

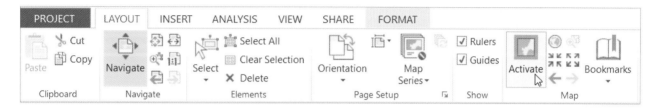

Note that the ribbon changes again, and you are now manipulating the map that resides within the map frame.

10) **Zoom in to the LA River and use the mouse to pan around the map.**

Note the map scale below the map is changing as you zoom in and out of the map.

11) **Use the LA River Layout Extent bookmark to return to the overview.**

While still in Activation mode, you want to reposition your LARiverBuffer more precisely in your layout to make room for your insert maps.

12) Use the zoom tools and map scale to position the buffer similar to the figure. This will give you room to the north and east of the buffer for the inset maps.

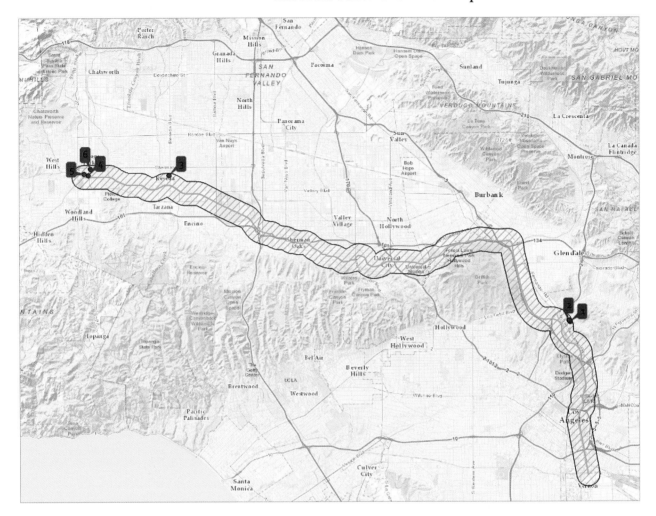

13) Go back to the layout by going to the Layout tab and, in the Map group, click Close Activation .

The ribbon resets for manipulating the layout.

Both sets of tools work in layout view, so you must be careful which navigation tools you are using.

Create a new map of recommended sites

1) Insert a new map and name it Lesson8 Recommended Sites.

2) Change the basemap to imagery. At the large scale of the inset map, imagery will show the site as it really looks. (Keep in mind that basemap layers are updated periodically, so your imagery may look different.)

3) In the Project pane, add the RecommendedSites feature class from the AnalysisOutputs project geodatabase.

4) Open the Symbology pane, and change the fill color to No Color, the Outline color to Solar Yellow, and the Outline width to 3 pt.

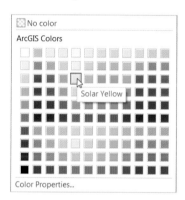

5) Click Apply.

Now you'll create bookmarks for each of the recommended sites to make it easier to add them to your layout.

6) Open the attribute table of the RecommendedSites layer.

7) In the attribute table, right-click the small gray square next to the first record (siteid1) and click Zoom To.

8) Create a bookmark and name it Site 1. This set of bookmarks will be saved under the Lesson8 Recommended Sites bookmarks.

9) Zoom to each of the following sites, 2–6, and make a bookmark of each site.

▶ Name each site by its site ID number.

10) Close the attribute table of the RecommendedSites layer.

Place inset maps and other graphic elements

An inset map is a small, supplementary map that shows part of the main map's area at a larger scale. It reveals detail not apparent at the scale of the main map. (An overview map, by contrast, shows the main map at a smaller scale to place it in the context of its surroundings.)

You want map readers to have a closeup look at each of the sites, so you'll create six large-scale inset maps. You can't simultaneously set different map scales within a single map frame (except with viewer windows), so you must insert some new map frames. In fact, you'll insert six of them and place them around the main map as shown in our layout sketch.

1) Open the Lesson8 layout [Lesson8 ×].

   ▸ The layout tab has a layout icon next to it, and the map tab has a map icon next to it.

You'll insert a new map frame to contain your first inset map.

2) On the Insert tab, in the Map Frames group, click the drop-down arrow on Map Frame. Scroll down the list of bookmarks to Lesson8 Recommended Sites and click Site 1.

3) Under Contents, open the properties of the new map frame, and under Options, rename it Site 1.

4) Under Placement, set Width and Height to 1.5. The values default to inches (in.).

Don't worry about the X and Y position values: you'll position the map frame manually.

5) Under the Map Frame drop-down arrow, click Border. Under Border properties, set the color to Gray 40%.

6) Change the Line width to 0.5 pt.

7) Click Apply and close the Format Map Frame pane.

8) On the Lesson8 layout, the Site 1 map frame is resized and renamed. If necessary, activate the map frame and zoom in so no other site boundaries show. Be sure to close activation if the map frame was activated.

When you add a basemap layer, the data providers are credited on the basemap itself. In data view, these credits appear as an icon in the lower-right corner of the map frame. (Clicking the icon opens the credits in a window.) In layout view, the credits are displayed as text. You may have noticed them before in the main map, but you can't miss them in the inset map, which is overwhelmed by the text. These credits can't be deleted, but they can be manipulated. In exercise 8c, you'll incorporate them into other data acknowledgments, but for the time being, you want to move them out of the way.

9) On the Insert tab, in the Text group, click Dynamic Text and then Service Layer Credits (near the bottom of the list).

Both sets of credits are moved out of their respective map frames and into a text box on the layout. The text box is selected.

10) Drag the text box anywhere outside the boundary of the layout page.

11) Click some empty white space to deselect the box.

With the service layer credits moved aside, you can turn your attention back to the inset map. Service layer credits are important because they provide credit to the basemap data providers. Therefore, you will add some data attribution later in the lesson.

Label the sites

You want to label the recommended sites with their Site IDs and areas so the map reader knows their ID number and how big they are. As you get closer to finishing the map, you'll be increasingly concerned with the exact placement and appearance of the text on the map.

1) Open the Lesson8 Recommended Sites map.

2) Open the RecommendedSites attribute table and, under the Menu button, click Fields View.

3) Change the numeric format of Acres to one decimal place by clicking Numeric and the Expand button [...] to open the Number Format pane.

4) Save your changes and close the Fields View table.

5) Confirm that the Acres field is displaying only one decimal place, and then close the RecommendedSites attribute table.

6) Select the RecommendedSites layer.

7) On the far left of the ribbon on the Labeling tab, in the Layer group, click Label to turn on labels.

8) In the Text Symbol group, change the font to Arial (or a similar font if not available) and the size to 8.

9) Set the color to Solar Yellow and the type to Bold.

10) Click the More Options button in the lower-right corner of the Text Symbol area of the ribbon.

11) Expand the Halo group and select the black rectangle from the Halo symbol drop-down menu. Then confirm that Halo size is 1 pt.

12) Click Apply. The labels now have a halo around them to help make them stand out against the outline of the parcel and basemap.

13) Switch to the Position tab by clicking Position at the top of the Label Class pane (next to the Symbol tab).

14) Under Placement, click Regular Placement, if necessary.

15) Click Straight in Polygon.

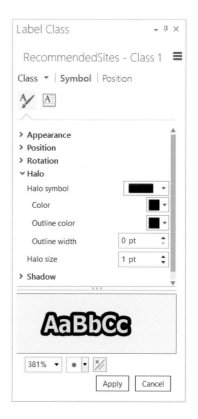

16) Select the check box next to May place label outside polygon boundary.

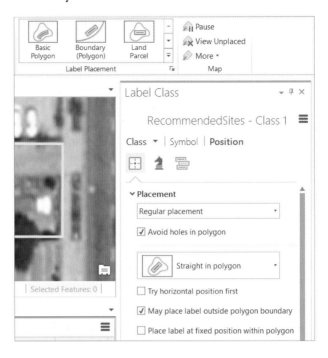

17) Close the Label Class pane.

18) In the Label Class group, click Expression .

A label expression lets you add custom text to an attribute value to make a more informative label.

19) In the Expression dialog box, for Language, click VBScript. Remove any existing text, and type the new expression as follows:
`"Site " & [SiteID] & vbnewline & [ACRES] & " Acres"`.

20) Check your expression against the figure.

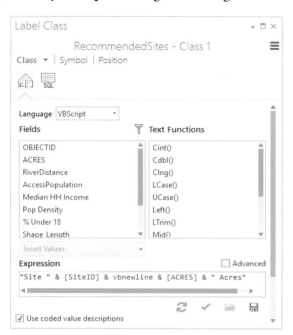

The expression tells ArcGIS Pro to take the word "Site" and tack on a space, and then add the value from the SiteID field. A new line is added, and then take the value from the ACRES field and tack on a space and then the word "Acres." Without the expression, the map reader wouldn't know what the units were.

21) Click Apply at the bottom of the Expression dialog box and then the Layer Properties dialog box.

The labeled feature should look similar to the figure.

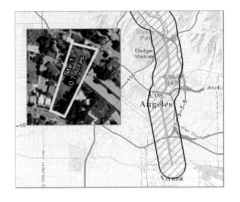

Since there isn't enough room to put all the inset maps right next to their sites, you'll move them to convenient locations, as shown in the layout sketch at the beginning of the lesson. Later, you'll connect them to the sites with leader lines.

22) On the Lesson8 layout, select the Site 1 map frame. (You should see the white selection handles.)

23) Drag the map frame to the lower-right corner of the main map, just to the left of the river buffer as shown in the figure.

Create the rest of the inset maps

1) On the Layout tab, in the Clipboard group on the far left, use the copy and paste functions to make a duplicate of the Site 1 map frame. The duplicate copy will be pasted right on top of the original Site 1 map frame in the map layout, so you'll need to move the copy off of Site 1.

2) Activate the duplicate of the Site 1 map frame you just moved by right-clicking on the map frame and clicking Activate.

3) On the Map tab, use Bookmarks to zoom to the Site 2 location.

The inset map zooms in to Site 2. In each inset map, you're letting the site boundary fill the map frame. That means the inset maps will be at different scales from one another—a fact you'll note on the map later on.

4) After confirming that the map frame is now zoomed to Site 2, on the Layout tab, in the Map group, click Close Activation.

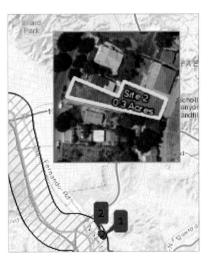

5) Under Contents, select Site 2, and on the layout, drag the Site 2 map frame to the approximate position shown in the figure, right above recommended sites 1 and 2.

6) Repeat the process, using steps 1–5, to create four more inset maps. Use the figure to guide your placement of the map frames. In summary, the steps are as follows:

- On the Layout tab, copy and paste the map frame you just made.
- Drag it to the appropriate place on the layout.
- Activate the new map frame.
- On the Map tab, use Bookmarks to zoom to the site location.
- On the Layout tab, click Close Activation.

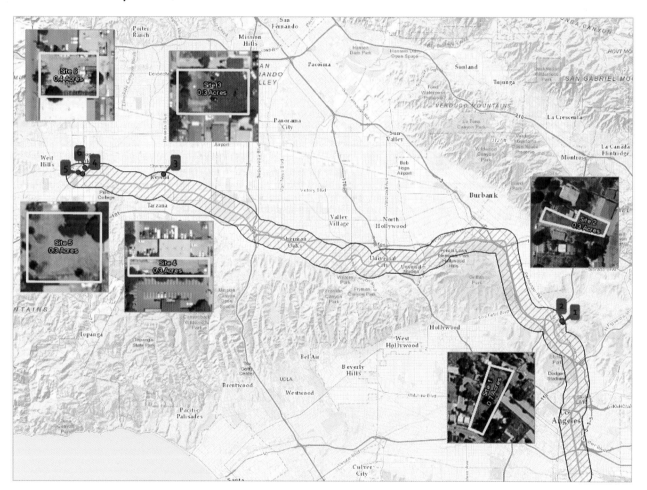

7) Under Contents, drag the inset map frames into sequential order.
8) When you're finished, save your project.

▶ Optionally, under Contents, collapse Lesson8.

Exercise 8c: Finish the map

The map still needs leader lines connecting the inset maps to their corresponding sites on the main map. Finally, you'll add a title, scale bar, north arrow, and acknowledgments to your map.

Add leader lines

You placed the inset maps away from the site locations, so you must connect them to their red symbols using leader lines.

1) If necessary, open the Lesson8 layout.

2) Zoom in on the Site 6 inset map and its corresponding symbol in the upper-left corner of the map layout.

3) On the Insert tab, in the Graphics group, confirm that Line is selected. Using the default line symbol drop-down arrow, select Limited Access Ramp under Scheme 3 to draw your lines.

4) Click on top of the Site 6 symbol to start the line. Go straight up; double-click on the edge of the inset map to complete the line.

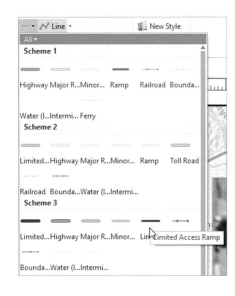

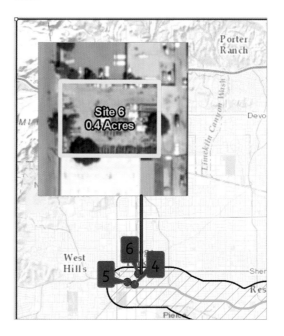

▶ To edit the completed leader line, right-click it to open its properties. In the Format Line Symbol pane, you can edit the color and line width under Symbol > Properties.

5) On the Layout tab, click the Full Extent button.

6) Zoom to another inset map, such as Site 4, and its corresponding symbol.

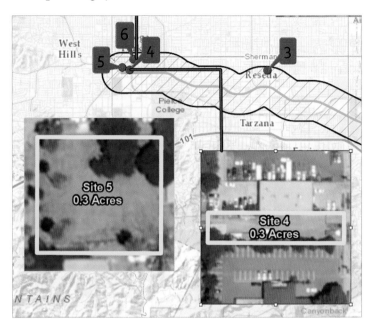

7) Use steps 4–6 to add leader lines for the remaining sites.

▶ If a line turns out badly, delete it by clicking Delete or Undo, and draw another one.

8) When you're finished, click the Full Extent button and deselect the last leader line.

9) Save your project.

Add a map title

You want the title to have the largest and most prominent type face on the layout.

1) On the Insert tab, in the Text group, click the Text drop-down arrow, and then click Rectangle to create a new rectangle paragraph for a title box.

2) Draw a box in any open space on the layout.

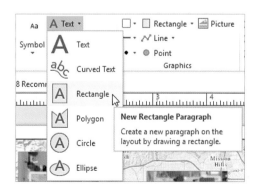

3) Open the properties of the Text element that was added to Contents.

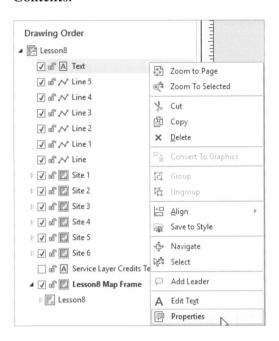

4) Expand General and change the name to Title.

5) Under Text, enter Potential LA River Park Site Locations in the text box. Delete the space between "River" and "Park," and then press Enter to add a line break. Click Apply.

6) On the Format ribbon, change the font to Arial and the size to 36 pt.

7) In the Current Selection group on the far left, change to the Background element.

8) Select a green for the background color.

9) Switch to Border and add a black line border.

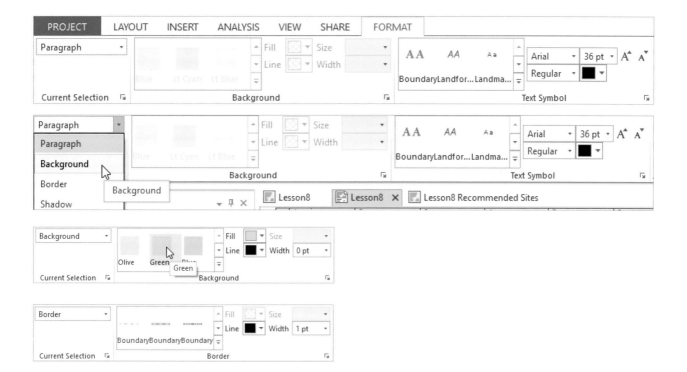

10) Resize the Title box and place in the upper-right corner of the layout. Then in the Contents pane, move the title to the top of the list.

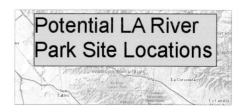

Add a Pacific Ocean label

You should identify the Pacific Ocean on the layout. The World Topographic Map does not label the ocean so you'll create your own text label. By convention, text for bodies of water is italicized.

1) Zoom and pan to the ocean area.

2) On the Insert tab, in the Text Group, click Curved Text.

3) Use the tool to create a curved path along the coastline. Add two to three points and double-click to end the line.

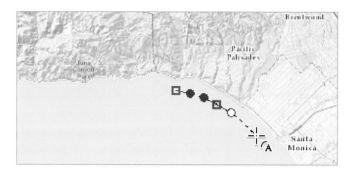

4) In the text box on the map, type Pacific Ocean.

5) On the Format ribbon, change the font to Arial with Italic. Change the font size to 12 pt, and choose a dark-blue color.

6) As you did above for the title, in the Contents pane open the properties of the text layer and rename it Pacific Ocean.

Add a river buffer label

You'll add one more label to clarify what the LA River buffer represents.

1) Zoom in to the area between the 405 and 101 Freeways (on the right of the Site 4 inset map).

2) Using steps 2 and 3 from the previous section, create a curved label just to the south of the buffer.

3) Type LA River 1/2 mile corridor.

4) On the Format ribbon, change the font to Arial with Bold. Change the font size to 12 pt, and choose a dark-green color.

5) In the Contents pane, open the properties of the text layer and rename it LA River buffer label.

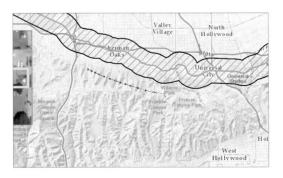 

Add a scale bar to the main map

A scale bar will give the map reader a sense of distance on the map.

1) On the Insert tab, in the Map Surrounds group, click the Scale Bar drop-down arrow.

In the drop-down menu, you can choose from a variety of scale bar types.

2) In the Imperial Group, select Single Division Scale Bar.

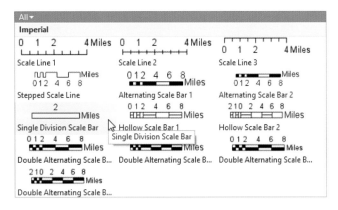

Exercise 8c: Finish the map   327

3) On the Scale Bar Design tab, make the following changes:
   - In the Divisions group, change Resize Behavior to Adjust width.
   - Change Division Value to 3.
   - Keep Divisions at 1 and Subdivisions at 0.
   - In the Numbers group, change Position to Center on bar.

4) On the Scale Bar Format tab, make the following changes:
   - Change the Fill color to Artic White.
   - Change the width to 0.5 pt.
   - Change the point size to 9 pt.

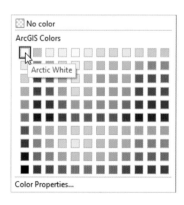

5) **Move the scale bar from the middle of the layout to a position below the Pacific Ocean label on the layout.**

> **Scale bars and scale text**
>
> Scale bars provide a visual indication of the size of features and the distance between features on the map. A scale bar is a line or bar divided into parts and labeled with its ground length, usually in multiples of map units such as tens of kilometers or hundreds of miles. If the map is enlarged or reduced, the scale bar remains correct. You can also represent the scale of your map with scale text, which is a verbal expression of scale. Scale text may relate equivalent units ("one inch equals two miles") or be a representative fraction that holds for any units (e.g., "1:100,000"). Scale text will be wrong if the map is enlarged or reduced after it has been printed.

Add a north arrow

You'll add a north arrow to remove any ambiguity about directions.

1) **On the Insert tab, click the North Arrow drop-down arrow to choose a north arrow.**

You need a slender arrow to economize on space in the corner of the map, but feel free to choose one you like.

2) **Select ArcGIS North 3.**

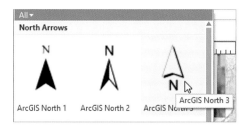

3) **On the North Arrow Format tab, change the size to 36 pt.**

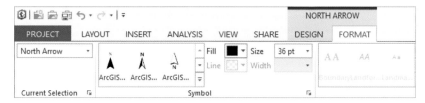

4) Move the north arrow from the middle of the layout to a position above your scale bar.

Add data attribution

It's important to acknowledge data sources. Credits for the World Topographic Map basemap layer are included in the service layer credits. You also want to credit the City and County of Los Angeles, which provided much of the project data. Finally, you want the map reader to know that the inset maps are at different scales. You'll combine all this information in one text element, which you'll work on in the next steps.

Earlier, in exercise 8b, you added dynamic text for the basemap credits and dragged them off the layout.

1) Drag the text box back and place it in the open space to the left of your north arrow and scale bar, and resize it so that it fits nicely. Not all the text will fit so you'll reformat the text.

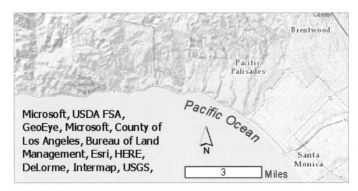

2) With the text box still selected, change the font to Arial Narrow 7 pt and Gray 50%.

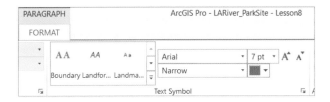

3) Double-click on the Layer Credit text box to open its properties.

The service layer credits are a type of dynamic text, which changes on the basis of the current properties of the map. If you had navigated to a different part of the world, the text displayed on the map layout would change to show the sources used on the basemap. Dynamic text uses tags, much like HTML, as placeholders for the text to be displayed and allows for the combination of dynamic and static text.

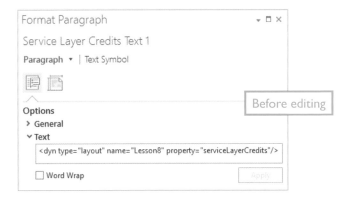

4) In the Format Paragraph dialog box, on the Paragraph tab, under Options > Text, position your cursor at the beginning of the line before the opening <dyn> tag, and enter the text Data Sources: with a colon at the end.

5) Position your pointer at the end of the line, and add two new lines by typing Shift + Enter twice.

6) Add the following text: Inset maps are at different scales.

7) Compare your edits to the figure, and then close the Paragraph box.

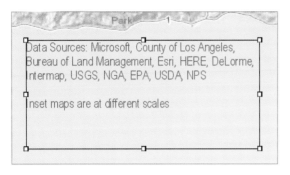

8) Click Apply.

9) Resize the text box and reposition it in the lower-left corner of the map to fit the space without crowding the other elements.

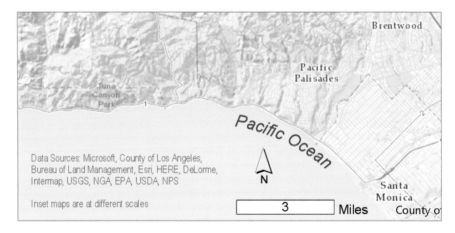

10) Save your project.

In the next section, you'll preview, export, and print the map. If you see any final adjustments you want to make, now is the time to do it. Look at the final map at the end of this lesson for reference.

Preview and share your map

If you have a printer available, you can print your map.

1) On the Layout tab, in the Navigate group, click the Full Extent button.

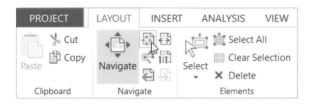

Your map should resemble the figure.

2) On the Share tab, in the Export group, click Layout.

3) Browse to your MapsAndMore folder and export your layout as a JPEG.

4) If you have access to a printer, you can also print from this same tab by clicking Layout 🖨 in the Print group.

5) Save your changes and close ArcGIS Pro.

Hopefully, your map looks good. The next, and final, phase of the project, in lesson 9, is to share your analysis results online with a web map.

Our goal in this book has been to guide you through the main steps of a GIS analysis problem: to explore the study area; to state the problem in quantitative terms; to choose, prepare, and edit the data; to plan and carry out the analysis; to automate the analysis with a model; and to present the analysis results. Not every GIS analysis problem is the same, but you should now understand the basic approach and have the requisite skills to take on the next project that comes your way.

Our second goal has been to introduce you to ArcGIS Pro software. In planning this book, we made a decision to use the software only insofar as it served the needs of the project. We tried to make the project complex enough that we would have reason to use a lot of functionality; nevertheless, we definitely didn't see or do everything. Keep exploring the software on your own, and visit the book's resource web page at esri.com/Understanding-GIS-3.

Lesson

9 Share your results online

THE CITY COUNCIL WOULD LIKE TO make the park analysis available to the public. A paper map, such as the one you created in lesson 8, is the traditional way to share results, but online maps are a new way to share your story, methods, and data with the widest possible audience. With ArcGIS Online, you can make web maps that combine Esri basemaps with your own geographic data and a huge number of map services (online layers) created by other agencies and users around the world. You can share your web maps with everyone or with user groups that you set up. You can also publish the maps as web apps to give them a self-contained and professional appearance.

You can use your ArcGIS Online account to store many kinds of geographic items in addition to web maps and layers. For example, you can upload geoprocessing packages, such as the one you created in lesson 7, that contain models, tools, and data. Likewise, you can upload projects, which bundle maps or individual layer files, along with their source data. These packages can be shared from your ArcGIS Online account, allowing you to communicate your maps, results, and workflows to others.

In the first exercise in this lesson, exercise 9a, you'll use ArcGIS Pro to prepare and publish web layers to ArcGIS Online. Next, in exercise 9b, you'll create a web map using these published layers, and in exercises 9c through 9f, you'll customize it, using Map Notes, bookmarks, and pop-up windows, to make it shine on the web. In the final two exercises, 9g and 9h, you'll use your web map to create two web apps: a Basic Viewer mapping app and an Esri® Story Map Journal℠ app, using simple "builders" without a single line of code. The Basic Viewer app will provide an easy-to-use app for users to explore your web map with simple tools for changing basemaps and layers, measuring, and printing. The Esri Story Map Journal template is ideal for combining narrative text and multimedia content along with your maps and geographic content to tell rich and engaging stories. You can download lesson 9 on the book's resource web page, at esri.com/Understanding-GIS-3.

Lesson Nine road map

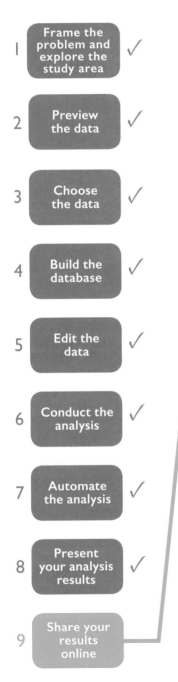
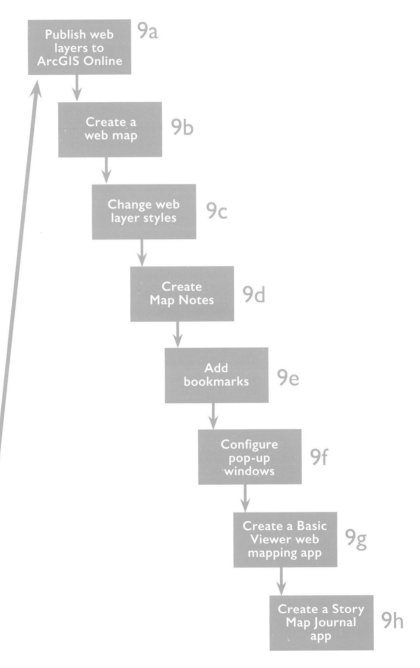

ArcGIS Online

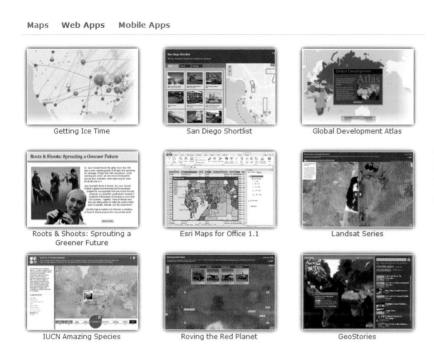

ArcGIS Online is GIS on the web. You can share your maps, data, apps, and workflows with everyone in the GIS community.

A selection of web apps that anyone can use from the growing ArcGIS Online gallery.

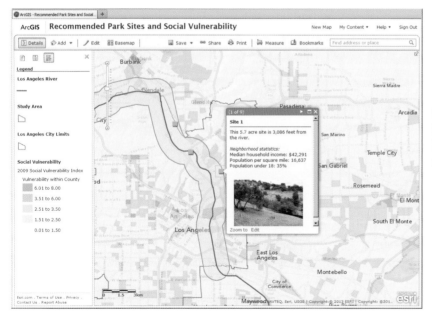

Create maps and apps from a web interface. Add your own data to Esri basemaps, and use online layers shared by Esri and others.

This web map shows the LA River park suitability results on top of a social vulnerability layer. The social vulnerability layer (comprising variables such as age, income, and disability) is a public map service on ArcGIS Online.

Introduction 337

ArcGIS Online (continued)

  ArcGIS Online and ArcGIS Desktop are part of the same system. Layers shared online can be added directly to ArcGIS Pro. Maps and analysis results created in ArcGIS Pro can be uploaded to ArcGIS Online.

The My Content page of your ArcGIS Online account is where you store, manage, and share items.

Items you can store include web maps, web apps, map services, data, and projects.

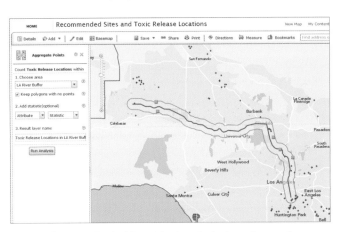

An ArcGIS Online subscription gives you even more capabilities. For example, you can publish your own map services and use online analysis and query tools.

This web map of suitable park sites includes a layer of facilities (red points) that release environmental toxins. An online analysis tool will count the number of toxic release locations in the buffer zone around the Los Angeles River.

Appendix

Data and image credits

Lesson 1

Imagery sources: Streets basemap—Esri, DeLorme, AND, Tele Atlas, First American, Esri Japan, UNEP-WCMC, USGS, METI, Esri Hong Kong, Esri Thailand, Procalculo Prosis; World Imagery basemap—Esri, DigitalGlobe, GeoEye, Earthstar Geographics, CNES/Airbus DS, USDA, USGS, AeroGRID, IGN, and the GIS User Community.

Data sources: US Populated Place Areas—from Esri Data & Maps (2009), courtesy of Tele Atlas, US Census Bureau; River—from Esri Data & Maps (2009), courtesy of USGS, Esri; US and Canada Parks—from Esri Data & Maps (2009), courtesy of Tele Atlas; US Tracts and Block Groups—Data and Maps for ArcGIS® (2015); LA County Land Parcels—Los Angeles GeoHub.

Lesson 2

Imagery sources: World Imagery basemap—Esri, DigitalGlobe, GeoEye, Earthstar Geographics, CNES/Airbus DS, USDA, USGS, AeroGRID, IGN, and the GIS User Community.

Data sources: LA County Land Parcels—Los Angeles GeoHub; US Populated Place Points—from Esri Data & Maps (2009), courtesy of US Census Bureau; Parks.shp—courtesy of LA City, Public Works, Bureau of Engineering; US Tracts and Block Groups—Data and Maps for ArcGIS (2015); US Block Centroids—courtesy of Tele Atlas North America Inc.

Lesson 3

Imagery sources: World Imagery basemap—Esri, DigitalGlobe, GeoEye, Earthstar Geographics, CNES/Airbus DS, USDA, USGS, AeroGRID, IGN, and the GIS User Community; Topographic basemap—Esri, HERE, DeLorme, Intermap, increment P Corp., GEBCO, USGS, FAO, NPS, NRCAN, GeoBase, IGN, Kadaster NL, Ordnance Survey, Esri Japan, METI, Esri China (Hong Kong), swisstopo, MapmyIndia, © OpenStreetMap contributors, and the GIS User Community.

Data sources: LARiver Centerline—courtesy of LA City, Public Works, Bureau of Engineering; US and Canada Water Polygons—from Esri Data & Maps (2009), courtesy of Tele Atlas; River—from Esri Data & Maps (2009), courtesy of USGS, Esri; US Populated Place Areas—from Esri Data & Maps (2009), courtesy of Tele Atlas, US Census Bureau; Parks.shp—courtesy of LA City, Public Works, Bureau of Engineering; US and Canada Parks—from Esri Data & Maps (2009), courtesy of Tele Atlas; US Counties—from Esri Data & Maps (2009), courtesy of ArcUSA, US Census Bureau, Esri (Pop2008 field); US States—from Esri Data & Maps (2009), courtesy of Tele Atlas; Continents—from Esri Data & Maps (2009), courtesy of *ArcWorld* Supplement.

Lesson 4

Imagery sources: Topographic basemap—Esri, HERE, DeLorme, Intermap, increment P Corp., GEBCO, USGS, FAO, NPS, NRCAN, GeoBase, IGN, Kadaster NL, Ordnance Survey, Esri Japan, METI, Esri China (Hong Kong), swisstopo, MapmyIndia, © OpenStreetMap contributors, and the GIS User Community; World Imagery basemap—Esri, DigitalGlobe, GeoEye, Earthstar Geographics, CNES/Airbus DS, USDA, USGS, AeroGRID, IGN, and the GIS User Community.

Data sources: LA County Land Parcels—Los Angeles GeoHub; US Highways—from Esri Data & Maps (2009), courtesy of Tele Atlas; US Counties—from Esri Data & Maps (2009), courtesy of ArcUSA, US Census Bureau, Esri (Pop2008 field); US Populated Place Areas—from Esri Data & Maps (2009), courtesy of Tele Atlas, US Census Bureau; US and Canada Parks—from Esri Data & Maps (2009), courtesy of Tele Atlas; US Tracts and Block Groups—Data and Maps for ArcGIS (2015).

Lesson 5

Imagery sources: World Imagery basemap—Esri, DigitalGlobe, GeoEye, Earthstar Geographics, CNES/Airbus DS, USDA, USGS, AeroGRID, IGN, and the GIS User Community.

Data sources: US and Canada Parks—from Esri Data & Maps (2009), courtesy of Tele Atlas.

Lesson 6

12 photos of LA City parcels: by Christian Harder.

Imagery sources: Topographic basemap—Esri, HERE, DeLorme, Intermap, increment P Corp., GEBCO, USGS, FAO, NPS, NRCAN, GeoBase, IGN, Kadaster NL, Ordnance Survey, Esri Japan, METI, Esri China (Hong Kong), swisstopo, MapmyIndia, © OpenStreetMap contributors, and the GIS User Community; World Imagery basemap—Esri, DigitalGlobe, GeoEye, Earthstar Geographics, CNES/Airbus DS, USDA, USGS, AeroGRID, IGN, and the GIS User Community.

Data sources: River—from Esri Data & Maps (2009), courtesy of USGS, Esri; US Populated Place Points—from Esri Data & Maps (2009), courtesy of US Census Bureau; US Counties—from Esri Data & Maps (2009), courtesy of ArcUSA, US Census Bureau, Esri (Pop2008 field); LARiver Centerline—courtesy of LA City, Public Works, Bureau of Engineering; US and Canada Parks—from Esri Data & Maps (2009), courtesy of Tele Atlas; LA County Land Parcels—Los Angeles GeoHub; US Block Groups—Data and Maps for ArcGIS (2015); US Block Centroids—courtesy of Tele Atlas North America Inc.; US Large Area Landmarks—from Esri Data & Maps (2009), courtesy of Tele Atlas; Los Angeles Metro Sherman Way Station in Canoga Park, California—Nearmap.

Lesson 7

Imagery sources: Topographic basemap—Esri, HERE, DeLorme, Intermap, increment P Corp., GEBCO, USGS, FAO, NPS, NRCAN, GeoBase, IGN, Kadaster NL, Ordnance Survey, Esri Japan, METI, Esri China (Hong Kong), swisstopo, MapmyIndia, © OpenStreetMap contributors, and the GIS User Community.

Data sources: LARiver Centerline—courtesy of LA City, Public Works, Bureau of Engineering; US Block Centroids—courtesy of Tele Atlas North America Inc.; US Block Groups—Data and Maps for ArcGIS (2015); US and Canada Parks—from Esri Data & Maps (2009), courtesy of Tele Atlas; LA County Land Parcels—Los Angeles GeoHub.

Lesson 8

Imagery sources: Topographic basemap—Esri, HERE, DeLorme, Intermap, increment P Corp., GEBCO, USGS, FAO, NPS, NRCAN, GeoBase, IGN, Kadaster NL, Ordnance Survey, Esri Japan, METI, Esri China (Hong Kong), swisstopo, MapmyIndia, © OpenStreetMap contributors, and the GIS User Community; World Imagery basemap—Esri, DigitalGlobe, GeoEye, Earthstar Geographics, CNES/Airbus DS, USDA, USGS, AeroGRID, IGN, and the GIS User Community.

Data sources: LARiver Centerline—courtesy of LA City, Public Works, Bureau of Engineering; LA County Land Parcels—Los Angeles GeoHub; US Populated Place Areas—from Esri Data & Maps (2009), courtesy of Tele Atlas, US Census Bureau; US Counties—from Esri Data & Maps (2009), courtesy of ArcUSA, US Census Bureau, Esri (Pop2008 field); US Highways—from Esri Data & Maps (2009), courtesy of Tele Atlas.

Lesson 9

Imagery sources: photo of LA City parcel: by Christian Harder. World Imagery basemap—Esri, DigitalGlobe, GeoEye, Earthstar Geographics, CNES/Airbus DS, USDA, USGS, AeroGRID, IGN, and the GIS User Community; Topographic basemap—Esri, HERE, DeLorme, Intermap, increment P Corp., GEBCO, USGS, FAO, NPS, NRCAN, GeoBase, IGN, Kadaster NL, Ordnance Survey, Esri Japan, METI, Esri China (Hong Kong), swisstopo, MapmyIndia, © OpenStreetMap contributors, and the GIS User Community.

Data sources: River—from Esri Data & Maps (2009), courtesy of USGS, Esri; US Median Household Income: © 2010 Esri.

Data license agreement

Downloadable data that accompanies this book is covered by a license agreement that stipulates the terms of use.

Index

A

age, children's, in park site selection, 80, 224
Alaska, Albers Equal Area Conic projection for, 118, 121–22
Albers Equal Area Conic coordinate system, 118, 121–22
analysis
   automating, 258–93. *See also* ModelBuilder, model(s), *and* model *entries*
   conducting, 203–55
      cleaning up map and geodatabase in, 241–49
      demographic constraints in, applying, 216–26
      proximity zones in, establishing, 206–16
      results of, evaluating, 249–55
      selecting suitable parcel in, 226–41
   essential tools for, 209–10
   plan for, 205–6
analysis results
   evaluating, 249–55
   presenting, 295–34
   sharing online, 335–38
ArcGIS Online, 335–38
ArcGIS Pro software
   adding data to, 185
   data acquisition from, 68
   editing tools of, 183
   learning aspects of, as goal, 100–102
   new map in, inserting, 9
   new project in, 8–9
   starting, 8, 36
area
   calculating, 180–81, 198–200
   measurement of, 114–15
attribute(s)
   common, in table joins, 172–74
   comparing, 155
   demographic, 224–26, 240–41. *See also* demographic attributes
   dissolving features by, 154
   software-managed, 75

attribute query(ies)
   explanation of, 28
   for demographic attributes, 224–26
   for subset selection, 228–29
   in model building, 275–77, 281–82
   in selecting Los Angeles River features, 25–27
attribute query problems, 203
attribute values
   apportioning, 223
   editing, 197–200
   in overlay operations, 219–22
   updating, 192–93
automating analysis, 258–93. *See also* ModelBuilder, model(s), *and* model *entries*

B

basemap(s)
   changing, 10
      to Imagery with Labels, 28–29
   dimming, 42
   layers of, 13. *See also* basemap layer(s)
   tiles of, 13
basemap layer(s), 13
   datasets and, 23
   definition of, 23
   features in, 14
   new features in, creating, 55
   of census tracts, 43–45. *See also* census tract layers
   of parks, adding, 37
   of rivers, adding, 23–24
   project data, adding, 13–14
   properties of, setting, 15–16
   saving as layer file, 31–33
block centroids
   adding to map, 236–38
   description of, 78
block_centroids shapefile, projecting to state plane coordinate system, 133–39
block group(s)
   demographic, assigning of, to proximity zone in model, 270–72

description of, 78
exporting to geodatabases, 163
identity overlays of, on proximity zone, 217–18
block group data, preparing, 161–71
bookmark(s)
creating, 29–30, 35, 252–54, 313
saving, 35
buffers
creating, 208, 211–13
definition of, 208
in models, 264–66, 268–70

C

calculation
of acreage, 180–81
of area, 198–200
of distances, 235–36
of fields, 168–70
of population density, 163–65
of summary statistics, 171
cartography, 295, 297. *See also* map(s)
census block, description of, 78
census block groups layer
adding, 49
scale range setting for, 51–52
symbolizing, by median household income, 49–51
census data, choosing, 104
census tract layer(s)
adding, 43
attribute table for, opening, 44
symbolizing, by population density, 44–45
census tracts
basemap layers of, 43–45. *See also* census tract layers
description of, 78
census units, previewing, 75–76
children, age of, in park site selection, 80, 224
cities, getting information on, 17–18
coordinate system(s)
changing, 124–25
in data preparation, 141
choosing, 106–25
geographic, 107, 118, 119. *See also* geographic coordinate systems
measurements and, 114–16
of map, changing, 116–17
of new map, changing, 141
of spatial datasets, determining, 108–10
on-the-fly projection of, 107–8
projected, 107, 118, 120
state plane, 123–24

copy features, 137, 139, 146–47
copying features, 146–47, 149–51, 159–60, 226, 230
in model building, 279
Create Thiessen Polygons tool, 209

D

data
acquiring, 68
adding to model, 263–65
block group, preparing, 161–71
choosing, 87–126
editing, 183–201
errors in, sources of, 183
examination of, 64–76
field, types of, 158
formats of, 129
input, storage of, 129
output, storage of, 129
parcel, choosing, 89
parks, adding, 98
preparing, 141–82. *See also* data preparation
previewing, 59–85
projecting, on the fly, 110–14
quantile classification method for, 46–47
representing real world as, 66–67
river
choosing, 91–92
symbolizing, 92–94
data analysis, exploratory, 36–38
data attribution, adding to map, 330–32
database(s), building, 127–71
data file geodatabase format, data in, 129
data preparation, 141–82
adding data to map in, 142
adding new map in, 141
changing coordinate system in, 141
hiding unnecessary field in, 144–45
reasons for, 141
data requirements
listing, 59, 61–64
table of, opening, 61
dataset(s), 87–106
choosing
considerations in, 87
for parcel data, 89
definition of, 23
layers and, 23
spatial, coordinate system of, determining, 108–10
datums, 126

definition query(ies)
   explanation of, 21, 28
   in filtering display of cities, 21–23
   in model building, 275–76
   on LA river, making, 24
demographic attributes
   adding, 240–41
   attribute queries for, 224–26
demographic block groups, assigning of, to proximity zone in model, 270–72
demographic constraints. See also age, children's; income; population density
   in conducting analysis, 216–26
dissolve operation, 98
dissolving features, 151–52, 154, 178–79
distance problems, 203
distances, calculating, 235–36

E
editing
   of existing features, 183, 185–93
   of new features, 193–200
   saving changes in, 192
   shortcut keys for, 187
   snapping behavior in, 190–91
environmental settings, 138–39
erasing features, 268–70
   in establishing proximity zones, 213–16

F
feature(s). See also feature class(es)
   copying, 226, 230
      in model building, 279
   creating, 195–96
   dissolving, 151–52, 154, 178–79
   erasing, 268–70
   existing, editing, 183, 185–93
   in basemap layers, 13, 14
      creating, 55
   new, editing, 193–200
   selecting, 142–44
   subselections, 228–29
feature class(es), 66
   copying, 138–40
   copying selected features to, 146–47, 149–51, 159–60
   creating, from selected features, 226
   geodatabase
      attributes of, 74
      field data types for, 158
   making layer from, 223
   shapefile, attributes of, 74
feature datasets, 65
feature identification (FID) attributes, software-managed, 74
feature templates, 194–95
field(s). See also attribute(s)
   adding, 245–46
   aliasing, 246, 248
   calculating, 168–70
   deleting, 243–45
   formatting, 246–48
   unnecessary
      hiding, 144–45
      turning off, 162–63
field data types, 158
Field Map of Join Features, 238
file geodatabase, description of, 130

G
geodatabase(s). See also feature class(es)
   ArcSDE® enterprise, 130
   cleaning up, 181–82, 244–48
   creating, 259–60
   default, 207
   exporting block groups to, 163
   for input and output data storage, 129
   results, creating, 206–7
   types of, 130
geographic coordinate systems, 107, 118, 119
   meridians in, 119
   parallels in, 119
geographic problem solving, framing problem in, 5–58. See also problem statement, framing
geography, US Census Bureau, 78
geoid, Earth as, 126
geometry, dissolving features by, 154
geoprocessing. See also geoprocessing tool(s)
   environmental settings in, 138–39
geoprocessing tool(s)
   Copy Features, 137, 139, 146–47, 149–51, 177, 226, 230, 279
   Create Thiessen Polygons, 209
   Dissolve, 151–52, 178–79
   Erase, 210, 213–16
   for projecting data to state plane coordinate system, 131–33
   Identity, 210, 240
   Make Feature Layer, 223, 275–78
   model as, 283–93

Near, 210, 235–36, 279–80
overlay, 210
Select Layer By Attribute, 210, 224
Select Layer By Location, 210, 227
Spatial Join, 210
GIS (geographic information system). *See also* ArcGIS Pro software
capabilities of, 4
graphic elements, placing, in map, 314–16

I
identity overlay(s), 240–41, 270–72
of block groups on proximity zone, 217–18
imagery
with labels, changing basemap to, 28–29
uses and features of, 28
income, median, in park site selection, 80–82, 84, 224
input data, storage of, 129

J
joins. *See* spatial join(s); table join(s)

L
label(s)
expression with, 318–20
Pacific Ocean, adding to map, 325–26
river buffer, adding to map, 326–27
scale range for, setting, 40–42
sites, 305–8, 316–20
labeling, of parks layer, 39–40
Lambert Conformal Conic map projection, 124
latitude, 119
layer(s)
adding descriptions to, 243
basemap, 13. *See also* basemap layer(s)
making from feature class, 223
saving as layer file, 249
symbolizing. *See* symbology
layer file, saving layer as, 31–33, 249
layout(s)
creating, 308–12
creating new map for, 298–99, 313
leader lines, adding to map, 322–23
longitude, 119
Los Angeles
city limits of, panning to, 35–36
record for, selecting, 19–20
symbology for, changing, 30–31

Los Angeles River
features of, selection of, query in, 25–27
flooding of, 1
following, 33–34
in growth of Los Angeles, 1
park near, finding suitable site for, 5, 7
perennial portion of, length of, finding, 27–28
revitalization of, 1–3
symbology of, 24–25
taming of, 1
Los Angeles River Revitalization Master Plan, 2–3

M
map(s)
adding block centroids to, 236–38
adding data to, 142
adding leader lines to, 322–23
adding map title to, 323–25
base. *See* basemap(s)
cities on
filtering display of, 21–23
getting information on, 17–18
cleaning up, 241–42
coordinate systems of. *See* coordinate system(s)
creating, 298–308, 313
definition of, 109
finishing, 322–34
functions of, 110
inset, placing, 314–16, 320–21
new, inserting, 9
overview, description of, 314
previewing and sharing, 333–34
zooming in on, 11–12
map frame, inserting, 310
map layouts. *See* layout(s)
mapping, thematic, 44
map projections, 120
Albers Equal Area Conic coordinate system, 118, 121–22
Bonne, 120
Eckert VI, 120
Fuller, 120
Lambert Conformal Conic, 124
Mercator, 120
Robinson, 120
Web Mercator coordinate system, 110–16, 117
Winkel Tripel, 120
map scale. *See also* scale bar(s)
in layout view, 311–12
setting, 33, 302

map title, adding to map, 323–25
measurement, area, 114–15
measuring distance, 48
Merge Rule, 239
meridians, in geographic coordinate systems, 119
metadata, 83
  definition of, 59
  in analysis of income attributes, 80–82
  official standards for, 83
model(s)
  adding data to, 263–65
  adding tool to, 264–67
  auto layout in, 269–70, 276, 279
  building, 263–83
  description of, 258
  outputs of, making tool parameters for, 289–90
  properties of, setting, 262
  queries in, 275–78
  reasons for, 258
  running, 272–75, 282
    as tool, 283–93. *See also* model tool
    vs. running model tool, 290
  setting up, 259–62
  understanding, 261
  validating, 272
ModelBuilder, 258. *See also* model(s) *and* model entries
model tool
  opening, 284
  parameters in, 284–87
  running, 287–89, 290–93
    versus running model, 290

N
Near tool, 210, 235–36, 279–80
neighborhood
  "densely populated," defining, 79
  with "lots of children," defining, 80
north arrow, adding to map, 329–30

O
on-the-fly projection
  of coordinate systems, 107–8
  of project data, 110–14
output data, storage of, 129
overlay(s)
  of block groups on proximity zone, 217–18
  identity, 270–72
overlay analysis, tools for, 210
overlay problems, 203

P
panning, along Los Angeles River, 33–36
parallels in geographic coordinate systems, 119
parameters for data management tools, 133
parcels
  previewing, 68–71
  vacant. *See* vacant parcels, suitable
park(s)
  data on
    choosing, 101–2
    symbolizing, 99
  examination of, in layers, 100–101
personal geodatabase, description of, 130
political boundaries, for final map, choosing, 104–5
population
  determining, 238–40
  park access, determining, in model building, 280–81
population density
  calculating, 163–65
  in park site selection, 84–85, 224
prime meridian, 119
problem statement
  framing, 5–58
    data analysis in, 7, 36–58
    data exploration in, 7, 36–58
    exploring study area in, 7–36. *See also* study area, exploring
    results of, 7
  reframing, 76–85
projected coordinate systems, 107, 118, 120
projecting data on the fly, 110–14
proximity tools, 209–10
proximity zone(s)
  assigning demographic block groups to, in model, 270–72
  establishing, 206–16
  identity overlays of block groups on, 217–18
pseudoprojection, geographic, 120

Q
quantile classification method for data, 46–47
quantitative symbology for census tract layer, 45–48
query(ies). *See also* attribute query(ies), definition query(ies), spatial query(ies)
  in models, 275–78
query tools, 209

R
raster data model, 67
records. *See also* feature(s)

reprojection, 120
river layer, adding, 23–24
rivers, basemap layers of, adding, 23–24
roads, for final map, choosing, 105–6

S

scale. *See* map scale
scale bar(s)
   adding to main map, 327–29
   scale text and, 329
scale range(s), 42
   for labels, setting, 40–42
   setting, 51–52
scale text, 329
Select By Location, 144
selecting features, 142–44
Select Layer By Attribute. *See* attribute query(ies)
   in parcel selection, 224, 228–29
Select Layer By Location. *See* spatial query(ies)
   in parcel selection, 227
Shape_Area attribute, software managed, 74
Shape_Area field, 163
shape attributes, software managed, 74
shapefile(s). *See also* feature class(es)
   projecting data in, to state plane coordinate system, 130–37
Shape_Length attribute, software managed, 74
snapping, 186, 190–91
SourceData folder, surveying, 65
spatial datasets, coordinate system of, determining, 108–10
spatial join(s)
   description of, 210, 236
   performing, 236
Spatial Join tool in model building, 280–81
spatial query(ies)
   explanation of, 28
   for vacant parcels in good zones in model building, 277–78
spheroid, Earth as, 126
state plane coordinate system, 123–24
   projecting data in shapefile to, 130–37
study area, exploring, 7–36
symbology
   for census block groups by median household income, 49–51
   for census tracts by population density, 44–45
   for layers with hatched fill, 299–300
   for Los Angeles, changing, 30–31
   for Los Angeles River, changing, 24–25
   for parks, 99
      changing, 37–38
   for recommended sites, 303–5
   for rivers, 92–95

T

table join(s)
   performing, 174, 176–77
table relates, 175
templates, feature, 194–95
thematic mapping, 44
tiles, basemap, 13
tool(s)
   adding to model, 264–67
   for analysis, essential, 209–10
   geoprocessing. *See* geoprocessing tool(s)
   proximity, 209–10
   query, 209
   running model as, 283–93. *See also* model tool
toolboxes, creating, 259
topological problems, 203

V

vacant parcels, suitable
   examining, 252–55
   inspecting, 230–35
   selection of, 227–41
vector data model, 66
vertices, editing, 187–92

W

Web Mercator coordinate system, 110–16, 117
workspaces, setting, 259–60
world
   continuous-surface view of, 67
   discrete-object view of, 66

Z

zooming
   in layout view, 311, 312
   to features, 11–12